Modern Chemistry®

Study Guide

HOLT, RINEHART AND WINSTON

A Harcourt Education Company

Orlando • **Austin** • New York • San Diego • Toronto • London

Contents

1 Matter and Change
Section 1 Chemistry Is a Physical Science. 1
Section 2 Matter and Its Properties 3
Section 3 Elements 5
Chapter 1 Mixed Review 7

2 Measurements and Calculations
Section 1 Scientific Method 9
Section 2 Units of Measurement 11
Section 3 Using Scientific Measurements 12
Chapter 2 Mixed Review 15

3 Atoms: The Building Blocks of Matter
Section 1 The Atom: From Philosophical Idea
           to Scientific Theory 17
Section 2 The Structure of the Atom. 19
Section 3 Counting Atoms 21
Chapter 3 Mixed Review 23

4 Arrangement of Electrons in Atoms
Section 1 The Development of a New Atomic Model 25
Section 2 The Quantum Model of the Atom 27
Section 3 Electron Configurations 29
Chapter 4 Mixed Review 31

5 The Periodic Law
Section 1 History of the Periodic Table 33
Section 2 Electron Configuration and the Periodic Table 35
Section 3 Electron Configuration and Periodic Properties 37
Chapter 5 Mixed Review 39

6 Chemical Bonding
Section 1 Introduction to Chemical Bonding 41
Section 2 Covalent Bonding and Molecular Compounds 43
Section 3 Ionic Bonding and Ionic Compounds 45
Section 4 Metallic Bonding. 47
Section 5 Molecular Geometry. 49
Chapter 6 Mixed Review 51

7 Chemical Formulas and Chemical Compounds
Section 1 Chemical Names and Formulas. 53
Section 2 Oxidation Numbers. 55
Section 3 Using Chemical Formulas 57
Section 4 Determining Chemical Formulas. 59
Chapter 7 Mixed Review 61

8 Chemical Equations and Reactions

Section 1 Describing Chemical Reactions.................... 65

Section 2 Types of Chemical Reactions..................... 67

Section 3 Activity Series of the Elements 69

Chapter 8 Mixed Review 71

9 Stoichiometry

Section 1 Introduction to Stoichiometry 73

Section 2 Ideal Stoichiometric Calculations 75

Section 3 Limiting Reactants and Percentage Yield 77

Chapter 9 Mixed Review 79

10 States of Matter

Section 1 Kinetic-Molecular Theory of Matter 81

Section 2 Liquids...................................... 83

Section 3 Solids....................................... 85

Section 4 Changes of State 87

Section 5 Water 89

Chapter 10 Mixed Review 91

11 Gases

Section 1 Gases and Pressure............................ 93

Section 2 The Gas Laws 95

Section 3 Gas Volumes and the Ideal Gas Law 97

Section 4 Diffusion and Effusion 99

Chapter 11 Mixed Review 100

12 Solutions

Section 1 Types of Mixtures 103

Section 2 The Solution Process 105

Section 3 Concentration of Solutions 107

Chapter 12 Mixed Review 109

13 Ions in Aqueous Solutions and Colligative Properties

Section 1 Compounds in Aqueous Solutions............... 111

Section 2 Colligative Properties of Solutions 113

Chapter 13 Mixed Review 115

14 Acids and Bases

Section 1 Properties of Acids and Bases.................. 117

Section 2 Acid-Base Theories........................... 119

Section 3 Acid-Base Reactions.......................... 121

Chapter 14 Mixed Review 123

15 Acid-Base Titration and pH

Section 1 Aqueous Solutions and the Concept of pH......... 125

Section 2 Determining pH and Titrations 127

Chapter 15 Mixed Review 129

16 Reaction Energy

    Section 1 Thermochemistry **131**

    Section 2 Driving Force of Reactions **133**

    Chapter 16 Mixed Review **135**

17 Reaction Kinetics

    Section 1 The Reaction Process **137**

    Section 2 Reaction Rates **139**

    Chapter 17 Mixed Review **141**

18 Chemical Equilibrium

    Section 1 The Nature of Chemical Equilibrium.............. **143**

    Section 2 Shifting Equilibrium **145**

    Section 3 Equilibria of Acids, Bases, and Salts............. **147**

    Section 4 Solubility Equilibrium.......................... **149**

    Chapter 18 Mixed Review **151**

19 Oxidation-Reduction Reactions

    Section 1 Oxidation and Reduction **153**

    Section 2 Balancing Redox Equations **155**

    Section 3 Oxidizing and Reducing Agents................. **157**

    Chapter 19 Mixed Review **159**

20 Electrochemistry

    Section 1 Introduction to Electrochemistry **161**

    Section 2 Voltaic Cells **163**

    Section 3 Electrolytic Cells **165**

    Chapter 20 Mixed Review **167**

21 Nuclear Chemistry

    Section 1 The Nucleus **169**

    Section 2 Radioactive Decay **171**

    Section 3 Nuclear Radiation............................. **173**

    Section 4 Nuclear Fission and Nuclear Fusion.............. **175**

    Chapter 21 Mixed Review **177**

22 Organic Chemistry

    Section 1 Organic Compounds........................... **179**

    Section 2 Hydrocarbons **181**

    Section 3 Functional Groups **183**

    Section 4 Organic Reactions **185**

    Chapter 22 Mixed Review **187**

23 Biological Chemistry

    Section 1 Carbohydrates and Lipids **189**

    Section 2 Amino Acids and Proteins **191**

    Section 3 Metabolism **193**

    Section 4 Nucleic Acids **195**

    Chapter 23 Mixed Review **197**

CHAPTER 1 REVIEW
Matter and Change

SHORT ANSWER   Answer the following questions in the space provided.

1. _____ Technological development of a chemical product often

   (a) lags behind basic research on the same substance.
   (b) does not involve chance discoveries.
   (c) is driven by curiosity.
   (d) is done for the sake of learning something new.

2. _____ The primary motivation behind basic research is to

   (a) develop new products.
   (b) make money.
   (c) understand an environmental problem.
   (d) gain knowledge.

3. _____ Applied research is designed to

   (a) solve a particular problem.
   (b) satisfy curiosity.
   (c) gain knowledge.
   (d) learn for the sake of learning.

4. _____ Chemistry is usually classified as

   (a) a biological science.
   (b) a physical science.
   (c) a social science.
   (d) a computer science.

5. Define the six major branches of chemistry.

_____

_____

_____

_____

_____

_____

_____

SECTION 1 continued

6. For each of the following types of chemical investigations, determine whether the investigation is *basic research, applied research,* or *technological development*. More than one choice may apply.

_____ a. A laboratory in a major university surveys all the reactions involving bromine.

_____ b. A pharmaceutical company explores a disease in order to produce a better medicine.

_____ c. A scientist investigates the cause of the ozone hole to find a way to stop the loss of the ozone layer.

_____ d. A pharmaceutical company discovers a more efficient method of producing a drug.

_____ e. A chemical company develops a new biodegradable plastic.

_____ f. A laboratory explores the use of ozone to inactivate bacteria in a drinking-water system.

7. Give examples of two different instruments routinely used in chemistry.

_____

_____

8. What are microstructures?

_____

_____

9. What is a chemical?

_____

_____

10. What is chemistry?

_____

_____

CHAPTER 1 REVIEW
Matter and Change

SECTION 2

SHORT ANSWER   Answer the following questions in the space provided.

1. Classify each of the following as a *homogeneous* or *heterogeneous* substance.

_____ **a.** iron ore

_____ **b.** quartz

_____ **c.** granite

_____ **d.** energy drink

_____ **e.** oil-and-vinegar salad dressing

_____ **f.** salt

_____ **g.** rainwater

_____ **h.** nitrogen

2. Classify each of the following as a *physical* or *chemical* change.

_____ **a.** ice melting

_____ **b.** paper burning

_____ **c.** metal rusting

_____ **d.** gas pressure increasing

_____ **e.** liquid evaporating

_____ **f.** food digesting

3. Compare a physical change with a chemical change.

_____

_____

_____

_____

_____

_____

SECTION 2 continued

4. Compare and contrast each of the following terms:

   a. *mass* and *matter*

   _____

   _____

   b. *atom* and *compound*

   _____

   _____

   _____

   c. *physical property* and *chemical property*

   _____

   _____

   _____

   _____

   _____

   d. *homogeneous mixture* and *heterogeneous mixture*

   _____

   _____

   _____

5. Using circles to represent particles, draw a diagram that compares the arrangement of particles in the solid, liquid, and gas states.

   Solid   Liquid   Gas

6. How is energy involved in chemical and physical changes?

   _____

   _____

   _____

   _____

CHAPTER 1 REVIEW
Matter and Change

SECTION 3

SHORT ANSWER Answer the following questions in the space provided.

1. A horizontal row of elements in the periodic table is called a(n) _____.

2. The symbol for the element in Period 2, Group 13, is _____.

3. Elements that are good conductors of heat and electricity are _____.

4. Elements that are poor conductors of heat and electricity are _____.

5. A vertical column of elements in the periodic table is called a(n) _____.

6. The ability of a substance to be hammered or rolled into thin sheets is called

   _____.

7. Is an element that is soft and easy to cut cleanly with a knife likely to be a metal or a

   nonmetal? _____

8. The elements in Group 18, which are generally unreactive, are called _____.

9. At room temperature, most metals are _____.

10. Name three characteristics of most nonmetals.

   _____

   _____

11. Name three characteristics of metals.

   _____

   _____

12. Name three characteristics of most metalloids.

   _____

   _____

13. Name two characteristics of noble gases.

   _____

SECTION 3 continued

14. What do elements of the same group in the periodic table have in common?

_____

_____

15. Within the same period of the periodic table, how do the properties of elements close to each other compare with the properties of elements far from each other?

_____

_____

_____

_____

16. You are trying to manufacture a new material, but you would like to replace one of the elements in your new substance with another element that has similar chemical properties. How would you use the periodic table to choose a likely substitute?

_____

_____

_____

17. What is the difference between a family of elements and elements in the same period?

_____

_____

_____

18. Complete the table below by filling in the spaces with correct names or symbols.

Name of element	Symbol of element
Aluminum	
	Ca
	Mn
Nickel	
Potassium	
Cobalt	
	Ag
	H

CHAPTER 1 REVIEW
Matter and Change

MIXED REVIEW

SHORT ANSWER Answer the following questions in the space provided.

1. Classify each of the following as a *homogeneous* or *heterogeneous* substance.

_____ **a.** sugar   _____ **d.** plastic wrap

_____ **b.** iron filings   _____ **e.** cement sidewalk

_____ **c.** granola bar

2. For each type of investigation, select the most appropriate branch of chemistry from the following choices: *organic chemistry, analytical chemistry, biochemistry, theoretical chemistry.* More than one branch may be appropriate.

_____ **a.** A forensic scientist uses chemistry to find information at the scene of a crime.

_____ **b.** A scientist uses a computer model to see how an enzyme will function.

_____ **c.** A professor explores the reactions that take place in a human liver.

_____ **d.** An oil company scientist tries to design a better gasoline.

_____ **e.** An anthropologist tries to find out the nature of a substance in a mummy's wrap.

_____ **f.** A pharmaceutical company examines the protein on the coating+ of a virus.

3. For each of the following types of chemical investigations, determine whether the investigation is *basic research, applied research,* or *technological development.* More than one choice may apply.

_____ **a.** A university plans to map all the genes on human chromosomes.

_____ **b.** A research team intends to find out why a lake remains polluted to try to find a way to clean it up.

_____ **c.** A science teacher looks for a solvent that will allow graffiti to be removed easily.

_____ **d.** A cancer research institute explores the chemistry of the cell.

_____ **e.** A professor explores the toxic compounds in marine animals.

MIXED REVIEW continued

4. Use the periodic table to identify the name, group number, and period number of the following elements:

_____ **a.** Cl

_____ **b.** Mg

_____ **c.** W

_____ **d.** Fe

_____ **e.** Sn

5. What is the difference between extensive and intensive properties?

_____

_____

6. Consider the burning of gasoline and the evaporation of gasoline. Which process represents a chemical change and which represents a physical change? Explain your answer.

_____

_____

_____

7. Describe the difference between a heterogeneous mixture and a homogeneous mixture, and give an example of each.

_____

_____

_____

8. Construct a concept map that includes the following terms: *atom, element, compound, pure substance, mixture, homogeneous,* and *heterogeneous*.

CHAPTER 2 REVIEW
Measurements and Calculations

SECTION 1

SHORT ANSWER   Answer the following questions in the space provided.

1. Determine whether each of the following is an example of *observation and data,* a *theory,* a *hypothesis,* a *control,* or a *model.*

_____ **a.** A research team records the rainfall in inches per day in a prescribed area of the rain forest. The square footage of vegetation and relative plant density per square foot are also measured.

_____ **b.** The intensity, duration, and time of day of the precipitation are noted for each precipitation episode. The types of vegetation in the area are recorded and classified.

_____ **c.** The information gathered is compared with the data on the average precipitation and the plant population collected over the last 10 years.

_____ **d.** The information gathered by the research team indicates that rainfall has decreased significantly. They propose that deforestation is the primary cause of this phenomenon.

2. "When 10.0 g of a white, crystalline sugar are dissolved in 100. mL of water, the solution is observed to freeze at $-0.54°C$, not $0.0°C$. The system is denser than pure water." Which parts of these statements represent quantitative information, and which parts represent qualitative information?

_____

_____

_____

_____

3. Compare and contrast a model with a theory.

_____

_____

_____

_____

SECTION 1 continued

4. Evaluate the models shown below. Describe how the models resemble the objects they represent and how they differ from the objects they represent.

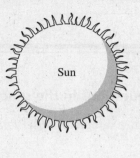

_____

_____

_____

_____

_____

_____

5. _____ How many different variables are represented in the two graphs shown below?

     a. one   **b.** two   **c.** three   **d.** four

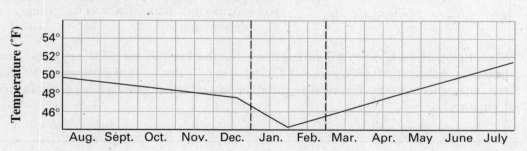

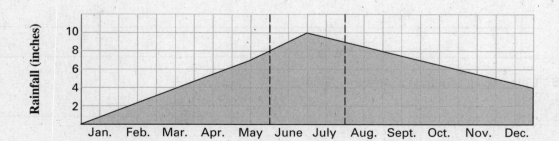

Name _____ Date _____ Class _____

SECTION 2

SHORT ANSWER Answer the following questions in the space provided.

1. Complete the following conversions:

    a. 100 mL = _____ L

    b. 0.25 g = _____ cg

    c. 400 cm^3 = _____ L

    d. 400 cm^3 = _____ m^3

2. For each measuring device shown below, identify the quantity measured and tell when it would remain constant and when it would vary.

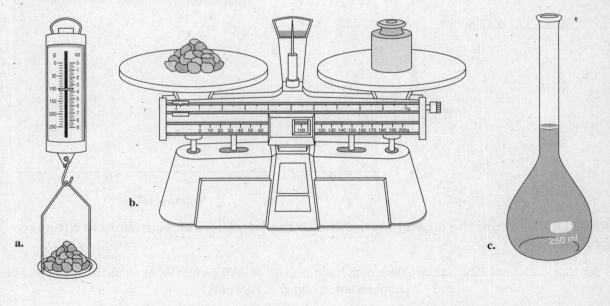

_____

_____

_____

_____

_____

_____

SECTION 2 continued

3. Use the data found in **Table 4** on page 38 of the text to answer the following questions:

_____ **a.** If ice were denser than liquid water at 0°C, would it float or sink in water?

_____ **b.** Water and kerosene do not dissolve readily in one another. If the two are mixed, they quickly separate into layers. Which liquid floats on top?

_____ **c.** The other liquids in **Table 4** that do not dissolve in water are gasoline, turpentine, and mercury. Which of these liquids would settle to the bottom when mixed with water?

4. Use the graph of the density of aluminum below to determine the approximate mass of aluminum samples with the following volumes.

_____ **a.** 8.0 mL

_____ **b.** 1.50 mL

_____ **c.** 7.25 mL

_____ **d.** 3.50 mL

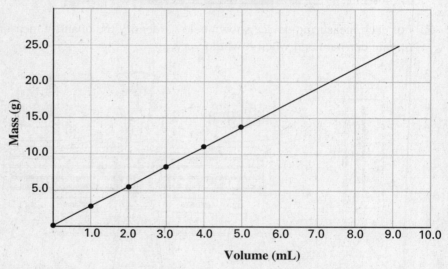

Mass vs. Volume of Aluminum

PROBLEMS Write the answer on the line to the left. Show all your work in the space provided.

5. _____ Aluminum has a density of 2.70 g/cm³. What would be the mass of a sample whose volume is 10.0 cm³?

6. _____ A certain piece of copper wire is determined to have a mass of 2.00 g per meter. How many centimeters of the wire would be needed to provide 0.28 g of copper?

CHAPTER 2 REVIEW
Measurements and Calculations

SECTION 3

SHORT ANSWER Answer the following questions in the space provided.

1. Report the number of significant figures in each of the following values:

   _____ **a.** 0.002 37 g   _____ **d.** 64 mL

   _____ **b.** 0.002 037 g   _____ **e.** 1.3×10^2 cm

   _____ **c.** 350. J   _____ **f.** 1.30×10^2 cm

2. Write the value of the following operations using scientific notation.

   _____ **a.** $\dfrac{10^3 \times 10^{-6}}{10^{-2}}$

   _____ **b.** $\dfrac{8 \times 10^3}{2 \times 10^5}$

   _____ **c.** $3 \times 10^3 + 4.0 \times 10^4$

3. The following data are given for two variables, A and B:

A	B
18	2
9	4
6	6
3	12

   a. In the graph provided, plot the data.

   _____ **b.** Are *A* and *B* directly or inversely proportional?

   _____ **c.** Do the data points form a straight line?

   _____ **d.** Which equation fits the relationship shown by the data?
   $\dfrac{A}{B} = k$ (a constant) or $A \times B = k$ (a constant)

   _____ **e.** What is the value of *k*?

SECTION 3 continued

4. Carry out the following calculations. Express each answer to the correct number of significant figures and use the proper units.

_____ a. 37.26 m + 2.7 m + 0.0015 m =

_____ b. 256.3 mL + 2 L + 137 mL =

_____ c. $\dfrac{300.\ kPa \times 274.57\ mL}{547\ kPa} =$

_____ d. $\dfrac{346\ mL \times 200\ K}{546.4\ K} =$

5. Round the following measurements to three significant figures.

_____ a. 22.77 g

_____ b. 14.62 m

_____ c. 9.3052 L

_____ d. 87.55 cm

_____ e. 30.25 g

PROBLEMS Write the answer on the line to the left. Show all your work in the space provided.

6. A pure solid at a fixed temperature has a constant density. We know that

$$density = \frac{mass}{volume}\ or\ D = \frac{m}{V}.$$

_____ a. Are mass and volume directly proportional or inversely proportional for a fixed density?

_____ b. If a solid has a density of 4.0 g/cm^3, what volume of the solid has a mass of 24 g?

7. A crime-scene tape has a width of 13.8 cm. A long strip of it is torn off and measured to be 56 m long.

_____ a. Convert 56 m into centimeters.

_____ b. What is the area of this rectangular strip of tape, in cm^2?

MODERN CHEMISTRY

CHAPTER 2 REVIEW

Measurements and Calculations

MIXED REVIEW

SHORT ANSWER Answer the following questions in the space provided.

1. Match the description on the right to the most appropriate quantity on the left.

_____ 2 m^3 **(a)** mass of a small paper clip

_____ 0.5 g   **(b)** length of a small paper clip

_____ 0.5 kg   **(c)** length of a stretch limousine

_____ 600 cm^2 **(d)** volume of a refrigerator compartment

_____ 20 mm   **(e)** surface area of the cover of this workbook

                                         (f) mass of a jar of peanut butter

2. _____ A measured quantity is said to have good accuracy if

(a) it agrees closely with the accepted value.
(b) repeated measurements agree closely.
(c) it has a small number of significant figures.
(d) all digits in the value are significant.

3. A certain sample with a mass of 4.00 g is found to have a volume of 7.0 mL. To calculate the density of the sample, a student entered 4.00 ÷ 7.0 on a calculator. The calculator display shows the answer as 0.571429.

_____ **a.** Is the setup for calculating density correct?

_____ **b.** How many significant figures should the answer contain?

4. It was shown in the text that in a value such as 4000 g, the precision of the number is uncertain. The zeros may or may not be significant.

_____ **a.** Suppose that the mass was determined to be 4000 g. How many significant figures are present in this measurement?

_____ **b.** Suppose you are told that the mass lies somewhere between 3950 and 4050 g. Use scientific notation to report the value, showing an appropriate number of significant figures.

5. If you divide a sample's mass by its density, what are the resulting units?

MIXED REVIEW continued

6. Three students were asked to determine the volume of a liquid by a method of their choosing. Each performed three trials. The table below shows the results. The actual volume of the liquid is 24.8 mL.

	Trial 1 (mL)	Trial 2 (mL)	Trial 3 (mL)
Student A	24.8	24.8	24.4
Student B	24.2	24.3	24.3
Student C	24.6	24.8	25.0

_____ **a.** Considering the average of all three trials, which student's measurements show the greatest accuracy?

_____ **b.** Which student's measurements show the greatest precision?

PROBLEMS Write the answer on the line to the left. Show all your work in the space provided.

7. _____ A single atom of platinum has a mass of 3.25×10^{-22} g. What is the mass of 6.0×10^{23} platinum atoms?

8. A sample thought to be pure lead occupies a volume of 15.0 mL and has a mass of 160.0 g.

_____ **a.** Determine its density.

_____ **b.** Is the sample pure lead? (Refer to **Table 4** on page 38 of the text.)

_____ **c.** Determine the percentage error, based on the accepted value for the density of lead.

MODERN CHEMISTRY

CHAPTER 3 REVIEW

Atoms: The Building Blocks of Matter

SECTION 1

SHORT ANSWER   Answer the following questions in the space provided.

1. Why is Democritus's view of matter considered only an idea, while Dalton's view is considered a theory?

_____

_____

_____

2. Give an example of a chemical or physical process that illustrates the law of conservation of mass.

_____

_____

_____

3. State two principles from Dalton's atomic theory that have been revised as new information has become available.

_____

_____

_____

4. The formation of water according to the equation

$$2H_2 + O_2 \rightarrow 2H_2O$$

shows that 2 molecules (made of 4 atoms) of hydrogen and 1 molecule (made of 2 atoms) of oxygen produce 2 molecules of water. The total mass of the product, water, is equal to the sum of the masses of each of the reactants, hydrogen and oxygen. What parts of Dalton's atomic theory are illustrated by this reaction? What law does this reaction illustrate?

_____

_____

_____

_____

_____

SECTION 1 continued

PROBLEMS Write the answer on the line to the left. Show all your work in the space provided.

5. _____ If 3 g of element C combine with 8 g of element D to form compound CD, how many grams of D are needed to form compound CD_2?

6. A sample of baking soda, $NaHCO_3$, *always* contains 27.37% by mass of sodium, 1.20% of hydrogen, 14.30% of carbon, and 57.14% of oxygen.

 a. Which law do these data illustrate?

 _____

 b. State the law.

 _____

 _____

 _____

 _____

 _____

 _____

7. Nitrogen and oxygen combine to form several compounds, as shown by the following table.

Compound	Mass of nitrogen that combines with 1 g oxygen (*g*)
NO	1.70
NO_2	0.85
NO_4	0.44

 Calculate the ratio of the masses of nitrogen in each of the following:

 _____ **a.** $\dfrac{NO}{NO_2}$ _____ **b.** $\dfrac{NO_2}{NO_4}$ _____ **c.** $\dfrac{NO}{NO_4}$

 d. Which law do these data illustrate?

 _____

18 ATOMS: THE BUILDING BLOCKS OF MATTER

MODERN CHEMISTRY

Atoms: The Building Blocks of Matter

SECTION 2

SHORT ANSWER   Answer the following questions in the space provided.

1. In cathode-ray tubes, the cathode ray is emitted from the negative electrode, which is called the

_____.

2. The smallest unit of an element that can exist either alone or in molecules containing the

same or different elements is the _____.

3. A positively charged particle found in the nucleus is called a(n) _____.

4. A nuclear particle that has no electrical charge is called a(n) _____.

5. The subatomic particles that are least massive and most massive, respectively, are the

_____ and _____.

6. A cathode ray produced in a gas-filled tube is deflected by a magnetic field. A wire carrying an
electric current can be pulled by a magnetic field. A cathode ray is deflected away from a
negatively charged object. What property of the cathode ray is shown by these phenomena?

_____

7. How would the electrons produced in a cathode-ray tube filled with neon gas compare with the
electrons produced in a cathode-ray tube filled with chlorine gas?

_____

_____

_____

8. a. Is an atom positively charged, negatively charged, or neutral?

_____

b. Explain how an atom can exist in this state.

_____

_____

_____

_____

SECTION 2 continued

9. Below are illustrations of two scientists' conceptions of the atom. Label the electrons in both illustrations with a − sign and the nucleus in the illustration to the right with a + sign. On the lines below the figures, identify which illustration was believed to be correct before Rutherford's gold-foil experiment and which was believed to be correct after Rutherford's gold-foil experiment.

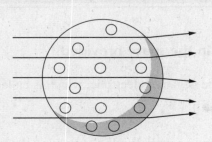   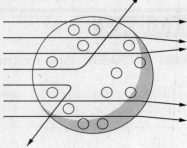

a. _____   b. _____

10. In the space provided, describe the locations of the subatomic particles in the labeled model of an atom of nitrogen below, and give the charge and relative mass of each particle.

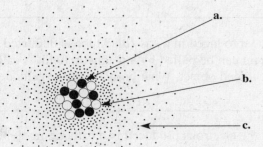

a. proton

_____

_____

b. neutron

_____

_____

c. electron (a possible location of this particle)

_____

_____

CHAPTER 3 REVIEW

Atoms: The Building Blocks of Matter

SECTION 3

SHORT ANSWER Answer the following questions in the space provided.

1. Explain the difference between the *mass number* and the *atomic number* of a nuclide.

_____

_____

_____

_____

2. Why is it necessary to use the average atomic mass of all isotopes, rather than the mass of the most commonly occurring isotope, when referring to the atomic mass of an element?

_____

_____

_____

_____

_____

3. How many particles are in 1 mol of carbon? 1 mol of lithium? 1 mol of eggs? Will 1 mol of each of these substances have the same mass?

_____

_____

_____

4. Explain what happens to each of the following as the atomic masses of the elements in the periodic table increase:

a. the number of protons

_____

b. the number of electrons

_____

c. the number of atoms in 1 mol of each element

_____

Name _____ Date _____ Class _____

SECTION 3 continued

5. Use a periodic table to complete the following chart:

Element	Symbol	Atomic number	Mass number
Europium-151			
	$^{109}_{47}Ag$		
Tellurium-128			

6. List the number of protons, neutrons, and electrons found in zinc-66.

_____ protons

_____ neutrons

_____ electrons

PROBLEMS Write the answer on the line to the left. Show all your work in the space provided.

7. _____ What is the mass in grams of 2.000 mol of oxygen atoms?

8. _____ How many moles of aluminum exist in 100.0 g of aluminum?

9. _____ How many atoms are in 80.45 g of magnesium?

10. _____ What is the mass in grams of 100 atoms of the carbon-12 isotope?

MODERN CHEMISTRY

CHAPTER 3 REVIEW

Atoms: The Building Blocks of Matter

MIXED REVIEW

SHORT ANSWER   Answer the following questions in the space provided.

1. The element boron, B, has an atomic mass of 10.81 amu according to the periodic table. However, no single atom of boron has a mass of exactly 10.81 amu. How can you explain this difference?

_____

_____

_____

2. How did the outcome of Rutherford's gold-foil experiment indicate the existence of a nucleus?

_____

_____

_____

_____

3. Ibuprofen, $C_{13}H_{18}O_2$, that is manufactured in Michigan contains 75.69% by mass carbon, 8.80% hydrogen, and 15.51% oxygen. If you buy some ibuprofen for a headache while you are on vacation in Germany, how do you know that it has the same percentage composition as the ibuprofen you buy at home?

_____

_____

_____

_____

4. Complete the following chart, using the atomic mass values from the periodic table:

Compound	Mass of Fe (g)	Mass of O (g)	Ratio of O:Fe
FeO			
Fe_2O_3			
Fe_3O_4			

Name _____ Date _____ Class _____

MIXED REVIEW continued

5. Complete the following table:

Element	Symbol	Atomic number	Mass number	Number of protons	Number of neutrons	Number of electrons
Sodium			22			
	F	9	19			
			80		45	
			40	20		
		1			0	
			222			86

PROBLEMS   Write the answer on the line to the left. Show all your work in the space provided.

6. _____ **a.** How many atoms are there in 2.50 mol of hydrogen?

_____ **b.** How many atoms are there in 2.50 mol of uranium?

_____ **c.** How many moles are present in 107 g of sodium?

7. A certain element exists as three natural isotopes, as shown in the table below.

Isotope	Mass (amu)	Percent natural abundance	Mass number
1	19.99244	90.51	20
2	20.99395	0.27	21
3	21.99138	9.22	22

_____ Calculate the average atomic mass of this element.

MODERN CHEMISTRY

CHAPTER 4 REVIEW

Arrangement of Electrons in Atoms

SECTION 1

SHORT ANSWER   Answer the following questions in the space provided.

1. In what way does the photoelectric effect support the particle theory of light?

_____

_____

_____

2. What is the difference between the ground state and the excited state of an atom?

_____

_____

3. Under what circumstances can an atom emit a photon?

_____

_____

4. How can the energy levels of the atom be determined by measuring the light emitted from an atom?

_____

_____

_____

_____

5. Why does electromagnetic radiation in the ultraviolet region represent a larger energy transition than does radiation in the infrared region?

_____

_____

_____

Name _____ Date _____ Class _____

SECTION 1 continued

6. Which of the waves shown below has the higher frequency? (The scale is the same for each drawing.) Explain your answer.

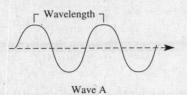

Wave A

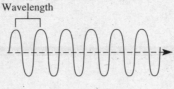

Wave B

_____

_____

_____

7. How many different photons of radiation were emitted from excited helium atoms to form the spectrum shown below? Explain your answer.

Spectrum for helium

_____

_____

_____

_____

_____

PROBLEMS Write the answer on the line to the left. Show all your work in the space provided.

8. _____ What is the frequency of light that has a wavelength of 310 nm?

9. _____ What is the wavelength of electromagnetic radiation if its frequency is 3.2×10^{-2} Hz?

Name _____ Date _____ Class _____

SECTION 2

SHORT ANSWER Answer the following questions in the space provided.

1. _____ How many quantum numbers are used to describe the properties of electrons in atomic orbitals?

    (a) 1   **(c)** 3

    (b) 2   **(d)** 4

2. _____ A spherical electron cloud surrounding an atomic nucleus would best represent

    (a) an *s* orbital.   **(c)** a combination of two different *p* orbitals.

    (b) a *p* orbital.   **(d)** a combination of an *s* and a *p* orbital.

3. _____ How many electrons can an energy level of $n = 4$ hold?

    (a) 32   **(c)** 8

    (b) 24   **(d)** 6

4. _____ How many electrons can an energy level of $n = 2$ hold?

    (a) 32   **(c)** 8

    (b) 24   **(d)** 6

5. _____ Compared with an electron for which $n = 2$, an electron for which $n = 4$ has more

    (a) spin.   **(c)** energy.

    (b) particle nature.   **(d)** wave nature.

6. _____ According to Bohr, which is the point in the figure below where electrons cannot reside?

    (a) point A   **(c)** point C

    (b) point B   **(d)** point D

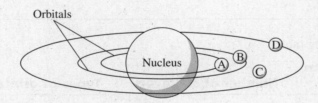

7. _____ According to the quantum theory, point D in the above figure represents

    (a) the fixed position of an electron.

    (b) the farthest position from the nucleus that an electron can achieve.

    (c) a position where an electron probably exists.

    (d) a position where an electron cannot exist.

SECTION 2 continued

8. How did de Broglie conclude that electrons have a wave nature?

_____

_____

_____

_____

_____

_____

9. Identify each of the four quantum numbers and the properties to which they refer.

_____

_____

_____

_____

_____

_____

_____

10. How did the Heisenberg uncertainty principle contribute to the idea that electrons occupy "clouds," or "orbitals"?

_____

_____

_____

_____

_____

_____

11. Complete the following table:

Principal quantum number, n	Number of sublevels	Types of orbitals
1		
2		
3		
4		

CHAPTER 4 REVIEW
Arrangement of Electrons in Atoms

SECTION 3

SHORT ANSWER   Answer the following questions in the space provided.

1. State the Pauli exclusion principle, and use it to explain why electrons in the same orbital must have opposite spin states.

   _____

   _____

   _____

   _____

   _____

2. Explain the conditions under which the following orbital notation for helium is possible:

   ↑   ↑
   _____   _____
     $1s$ $2s$

   _____

   _____

Write the ground-state electron configuration and orbital notation for each of the following atoms:

3. Phosphorus

4. Nitrogen

5. Potassium

SECTION 3 continued

6. Aluminum

7. Argon

8. Boron

9. Which guideline, Hund's rule or the Pauli exclusion principle, is violated in the following orbital diagrams?

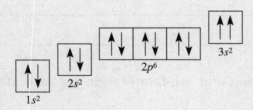

a. _____

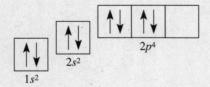

b. _____

CHAPTER 4 REVIEW

Arrangement of Electrons in Atoms

MIXED REVIEW

SHORT ANSWER   Answer the following questions in the space provided.

1. Under what conditions is a photon emitted from an atom?

_____

_____

_____

2. What do quantum numbers describe?

_____

_____

_____

3. What is the relationship between the principal quantum number and the electron configuration?

_____

_____

_____

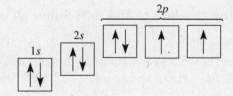

4. In what way does the figure above illustrate Hund's rule?

_____

_____

_____

5. In what way does the figure above illustrate the Pauli exclusion principle?

_____

_____

MIXED REVIEW continued

6. Elements of the fourth and higher main-energy levels do not seem to follow the normal sequence for filling orbitals. Why is this so?

_____

_____

_____

7. How do electrons create the colors in a line-emission spectrum?

_____

_____

_____

8. Write the ground-state electron configuration of the following atoms:

a. Carbon

_____

b. Potassium

_____

c. Gallium

_____

d. Copper

_____

PROBLEMS   **Write the answer on the line to the left. Show all your work in the space provided.**

9. _____ What is the wavelength of light that has a frequency of 3×10^{-4} Hz in a vacuum?

10. _____ What is the energy of a photon that has a frequency of 5.0×10^{14} Hz?

CHAPTER 5 REVIEW
The Periodic Law

SECTION 1

SHORT ANSWER   Answer the following questions in the space provided.

1. _____ In the modern periodic table, elements are ordered

   (a) according to decreasing atomic mass.
   (b) according to Mendeleev's original design.
   (c) according to increasing atomic number.
   (d) based on when they were discovered.

2. _____ Mendeleev noticed that certain similarities in the chemical properties of elements appeared at regular intervals when the elements were arranged in order of increasing

   (a) density.   **(c)** atomic number.
   (b) reactivity.   **(d)** atomic mass.

3. _____ The modern periodic law states that

   (a) no two electrons with the same spin can be found in the same place in an atom.
   (b) the physical and chemical properties of an element are functions of its atomic number.
   (c) electrons exhibit properties of both particles and waves.
   (d) the chemical properties of elements can be grouped according to periodicity, but physical properties cannot.

4. _____ The discovery of the noble gases changed Mendeleev's periodic table by adding a new

   (a) period.   **(c)** group.
   (b) series.   **(d)** level.

5. _____ The most distinctive property of the noble gases is that they are

   (a) metallic.   **(c)** metalloid.
   (b) radioactive.   **(d)** largely unreactive.

6. _____ Lithium, the first element in Group 1, has an atomic number of 3. The second element in this group has an atomic number of

   (a) 4.   **(c)** 11.
   (b) 10.   **(d)** 18.

7. An isotope of fluorine has a mass number of 19 and an atomic number of 9.

   _____ **a.** How many protons are in this atom?

   _____ **b.** How many neutrons are in this atom?

   _____ **c.** What is the nuclear symbol of this fluorine atom, including its mass number and atomic number?

SECTION 1 continued

8. Samarium, Sm, is a member of the lanthanide series.

_____ **a.** Identify the element just below samarium in the periodic table.

_____ **b.** By how many units do the atomic numbers of these two elements differ?

9. A certain isotope contains 53 protons, 78 neutrons, and 54 electrons.

_____ **a.** What is its atomic number?

_____ **b.** What is the mass number of this atom?

_____ **c.** What is the name of this element?

_____ **d.** Identify two other elements that are in the same group as this element.

10. In a modern periodic table, every element is a member of both a horizontal row and a vertical column. Which one is the group, and which one is the period?

_____

_____

11. Explain the distinction between atomic mass and atomic number of an element.

_____

_____

_____

_____

12. In the periodic table, the atomic number of I is greater than that of Te, but its atomic mass is less. This phenomenon also occurs with other neighboring elements in the periodic table. Name two of these pairs of elements. Refer to the periodic table if necessary.

_____

_____

_____

_____

CHAPTER 5 REVIEW
The Periodic Law

SECTION 2

SHORT ANSWER   Use this periodic table to answer the following questions in the space provided.

1. Identify the element and write the noble-gas notation for each of the following:

a. the Group 14 element in Period 4

_____

b. the only metal in Group 15

_____

c. the transition metal with the smallest atomic mass

_____

d. the alkaline-earth metal with the largest atomic number

_____

SECTION 2 continued

2. On the periodic table given, several areas are labeled with letters A–H.

_____ **a.** Which block does **A** represent, *s, p, d,* or *f*?

b. Identify the remaining labeled areas of the table, choosing from the following terms: *main-group elements, transition elements, lanthanides, actinides, alkali metals, alkaline-earth metals, halogens, noble gases.*

_____ **B**

_____ **C**

_____ **D**

_____ **E**

_____ **F**

_____ **G**

_____ **H**

3. Give the symbol, period, group, and block for the following:

a. sulfur

_____

_____

b. nickel

_____

_____

c. $[Kr]5s^1$

_____

_____

d. $[Ar]3d^54s^1$

_____

_____

4. There are 18 columns in the periodic table; each has a group number. Give the group numbers that make up each of the following blocks:

_____ **a.** *s* block

_____ **b.** *p* block

_____ **c.** *d* block

Name _____ Date _____ Class _____

CHAPTER 5 REVIEW
The Periodic Law

SECTION 3

SHORT ANSWER Answer the following questions in the space provided.

1. _____ When an electron is added to a neutral atom, energy is

    (a) always absorbed.   **(c)** either absorbed or released.

    (b) always released.   **(d)** neither absorbed nor released.

2. _____ The energy required to remove an electron from a neutral atom is the atom's

    (a) electron affinity.   **(c)** electronegativity.

    (b) electron energy.   **(d)** neither absorbed nor released.

3. From left to right across a period on the periodic table,

_____ **a.** electron affinity values tend to become more (negative or positive).

_____ **b.** ionization energy values tend to (increase or decrease).

_____ **c.** atomic radii tend to become (larger or smaller).

4. _____ **a.** Name the halogen with the least-negative electron affinity.

     _____ **b.** Name the alkali metal with the highest ionization energy.

     _____ **c.** Name the element in Period 3 with the smallest atomic radius.

     _____ **d.** Name the Group 14 element with the largest electronegativity.

5. Write the electron configuration of the following:

a. Na

_____

b. Na^+

_____

c. O

_____

d. O^{2-}

_____

e. Co^{2+}

_____

SECTION 3 continued

6. a. Compare the radius of a positive ion to the radius of its neutral atom.

_____

_____

b. Compare the radius of a negative ion to the radius of its neutral atom.

_____

_____

7. a. Give the approximate positions and blocks where metals and nonmetals are found in the periodic table.

_____

_____

_____

b. Of metals and nonmetals, which tend to form positive ions? Which tend to form negative ions?

_____

_____

8. Table 3 on page 155 of the text lists successive ionization energies for several elements.

_____ **a.** Identify the electron that is removed in the first ionization energy of Mg.

_____ **b.** Identify the electron that is removed in the second ionization energy of Mg.

_____ **c.** Identify the electron that is removed in the third ionization energy of Mg.

d. Explain why the second ionization energy is higher than the first, the third is higher than the second, and so on.

_____

_____

_____

_____

9. Explain the role of valence electrons in the formation of chemical compounds.

_____

_____

_____

_____

CHAPTER 5 REVIEW
The Periodic Law

MIXED REVIEW

SHORT ANSWER   Answer the following questions in the space provided.

1. Consider the neutral atom with 53 protons and 74 neutrons to answer the following questions.

_____ **a.** What is its atomic number?

_____ **b.** What is its mass number?

_____ **c.** Is the element's position in a modern periodic table determined by its atomic number or by its atomic mass?

2. Consider an element whose outermost electron configuration is $3d^{10}4s^24p^x$.

_____ **a.** To which period does the element belong?

_____ **b.** If it is a halogen, what is the value of x?

_____ **c.** The group number will equal $(10 + 2 + x)$. True or False?

3. _____ **a.** In which block are metalloids found, s, p, d, or f?

_____ **b.** In which block are the hardest, densest metals found, s, p, or d?

4. _____ **a.** Name the most chemically active halogen.

_____ **b.** Write its electron configuration.

_____ **c.** Write the configuration of the most stable ion this element makes.

5. Refer only to the periodic table at the top of the review of Section 2 to answer the following questions on periodic trends.

_____ **a.** Which has the larger radius, Al or In?

_____ **b.** Which has the larger radius, Se or Ca?

_____ **c.** Which has a larger radius, Ca or Ca^{2+}?

_____ **d.** Which class has greater ionization energies, metals or nonmetals?

_____ **e.** Which has the greater ionization energy, As or Cl?

_____ **f.** An element with a large negative electron affinity is most likely to form a (positive ion, negative ion, or neutral atom)?

Name _____ Date _____ Class _____

MIXED REVIEW continued

_____ **g.** In general, which has a stronger electron attraction, a large atom or a small atom?

_____ **h.** Which has greater electronegativity, O or Se?

_____ **i.** In the covalent bond between Se and O, to which atom is the electron pair more closely drawn?

_____ **j.** How many valence electrons are there in a neutral atom of Se?

6. _____ Identify all of the following ions that do not have noble-gas stability. K^+ S^{2-} Ca^+ I^- Al^{3+} Zn^{2+}

7. Use only the periodic table in the review of Section 2 to give the noble-gas notation of the following:

_____ **a.** Br

_____ **b.** Br^-

_____ **c.** the element in Group 13, Period 5

_____ **d.** the lanthanide with the smallest atomic number

8. Use electron configuration and position in the periodic table to describe the chemical properties of calcium and oxygen.

_____

_____

_____

_____

_____

_____

9. Copper's electron configuration might be predicted to be $3d^9 4s^2$. But in fact, its configuration is $3d^{10} 4s^1$. The two elements below copper in Group 11 behave similarly. (Confirm this in the periodic table in **Figure 6** on pages 140–141 of the text.)

_____ **a.** Which configuration for copper is apparently more stable?

_____ **b.** Is the d sublevel completed in the atoms of these three elements?

_____ **c.** Every element in Period 4 has four levels of electrons established. True or False?

CHAPTER 6 REVIEW
Chemical Bonding

SECTION 1

SHORT ANSWER   Answer the following questions in the space provided.

1. _____ A chemical bond between atoms results from the attraction between the valence electrons and _____ of different atoms.

   (a) nuclei
   (b) inner electrons
   (c) isotopes
   (d) Lewis structures

2. _____ A covalent bond consists of

   (a) a shared electron.
   (b) a shared electron pair.
   (c) two different ions.
   (d) an octet of electrons.

3. _____ If two covalently bonded atoms are identical, the bond is identified as

   (a) nonpolar covalent.
   (b) polar covalent.
   (c) ionic.
   (d) dipolar.

4. _____ A covalent bond in which there is an unequal attraction for the shared electrons is

   (a) nonpolar.
   (b) polar.
   (c) ionic.
   (d) dipolar.

5. _____ Atoms with a strong attraction for electrons they share with another atom exhibit

   (a) zero electronegativity.
   (b) low electronegativity.
   (c) high electronegativity.
   (d) Lewis electronegativity.

6. _____ Bonds that possess between 5% and 50% ionic character are considered to be

   (a) ionic.
   (b) pure covalent.
   (c) polar covalent.
   (d) nonpolar covalent.

7. _____ The greater the electronegativity difference between two atoms bonded together, the greater the bond's percentage of

   (a) ionic character.
   (b) nonpolar character.
   (c) metallic character.
   (d) electron sharing.

8. The electrons involved in the formation of a chemical bond are called

   _____.

9. A chemical bond that results from the electrostatic attraction between positive and

   negative ions is called a(n) _____.

SECTION 1 continued

10. If electrons involved in bonding spend most of the time closer to one atom rather than

the other, the bond is _____.

11. If a bond's character is more than 50% ionic, then the bond is called

a(n) _____.

12. A bond's character is more than 50% ionic if the electronegativity difference between the two

atoms is greater than _____.

13. Write the formula for an example of each of the following compounds:

_____ **a.** nonpolar covalent compound

_____ **b.** polar covalent compound

_____ **c.** ionic compound

14. Describe how a covalent bond holds two atoms together.

_____

_____

_____

15. What property of the two atoms in a covalent bond determines whether or not the bond will be
polar?

_____

_____

16. How can electronegativity be used to distinguish between an ionic bond and a covalent bond?

_____

_____

_____

_____

17. Describe the electron distribution in a polar-covalent bond and its effect on the partial charges of
the compound.

_____

_____

_____

CHAPTER 6 REVIEW
Chemical Bonding

SECTION 2

SHORT ANSWER   Answer the following questions in the space provided.

1. Use the concept of potential energy to describe how a covalent bond forms between two atoms.

_____

_____

_____

_____

_____

_____

2. Name two elements that form compounds that can be exceptions to the octet rule.

_____

_____

3. Explain why resonance structures are used instead of Lewis structures to correctly model certain molecules.

_____

_____

_____

_____

4. Bond energy is related to bond length. Use the data in the tables below to arrange the bonds listed in order of increasing bond length, from shortest bond to longest.

a.

Bond	Bond energy (kJ/mol)
H—F	569
H—I	299
H—Cl	432
H—Br	366

_____

b.

Bond	Bond energy (kJ/mol)
C—C	346
C≡C	835
C=C	612

_____

_____

5. Draw Lewis structures to represent each of the following formulas:

a. NH_3

b. H_2O

c. CH_4

d. C_2H_2

e. CH_2O

CHAPTER 6 REVIEW
Chemical Bonding

SECTION 3

SHORT ANSWER Answer the following questions in the space provided.

1. _____ The notation for sodium chloride, NaCl, stands for one

  (a) formula unit.   (c) crystal.
  (b) molecule.   (d) atom.

2. _____ In a crystal of an ionic compound, each cation is surrounded by a number of

  (a) molecules.   (c) dipoles.
  (b) positive ions.   (d) negative ions.

3. _____ Compared with the neutral atoms involved in the formation of an ionic compound, the crystal lattice that results is

  (a) higher in potential energy.   (c) equal in potential energy.
  (b) lower in potential energy.   (d) unstable.

4. _____ The lattice energy of compound A is greater in magnitude than that of compound B. What can be concluded from this fact?

  (a) Compound A is not an ionic compound.
  (b) It will be more difficult to break the bonds in compound A than those in compound B.
  (c) Compound B has larger crystals than compound A.
  (d) Compound A has larger crystals than compound B.

5. _____ The forces of attraction between molecules in a molecular compound are generally

  (a) stronger than the attractive forces among formula units in ionic bonding.
  (b) weaker than the attractive forces among formula units in ionic bonding.
  (c) approximately equal to the attractive forces among formula units in ionic bonding.
  (d) equal to zero.

6. Describe the force that holds two ions together in an ionic bond.

_____

_____

_____

7. What type of energy best represents the strength of an ionic bond?

_____

_____

SECTION 3 continued

8. What types of bonds are present in an ionic compound that contains a polyatomic ion?

_____

_____

_____

_____

9. Arrange the ionic bonds in the table below in order of increasing strength from weakest to strongest.

Ionic bond	Lattice energy (kJ/mol)
NaCl	−787
CaO	−3384
KCl	−715
MgO	−3760
LiCl	−861

_____

_____

10. Draw Lewis structures for the following polyatomic ions:

a. NH_4^+

b. SO_4^{2-}

11. Draw the two resonance structures for the nitrite anion, NO_2^-.

CHAPTER 6 REVIEW
Chemical Bonding

SECTION 4

SHORT ANSWER   Answer the following questions in the space provided.

1. _____ In metals, the valence electrons are considered to be

    (a) attached to particular positive ions.   **(c)** immobile.
    (b) shared by all surrounding atoms.   **(d)** involved in covalent bonds.

2. _____ The fact that metals are malleable and ionic crystals are brittle is best explained in terms of their

    (a) chemical bonds.   **(c)** enthalpies of vaporization.
    (b) London forces.   **(d)** polarity.

3. _____ As light strikes the surface of a metal, the electrons in the electron sea

    (a) allow the light to pass through.
    (b) become attached to particular positive ions.
    (c) fall to lower energy levels.
    (d) absorb and re-emit the light.

4. _____ Mobile electrons in the metallic bond are responsible for

    (a) luster.   **(c)** electrical conductivity.
    (b) thermal conductivity.   **(d)** All of the above.

5. _____ In general, the strength of the metallic bond _____ moving from left to right on any row of the periodic table.

    (a) increases   **(c)** remains the same
    (b) decreases   **(d)** varies

6. _____ When a metal is drawn into a wire, the metallic bonds

    (a) break easily.   **(c)** do not break.
    (b) break with difficulty.   **(d)** become ionic bonds.

7. Use the concept of electron configurations to explain why the number of valence electrons in metals tends to be less than the number in most nonmetals.

_____

_____

_____

_____

SECTION 4 continued

8. How does the behavior of electrons in metals contribute to the metal's ability to conduct electricity and heat?

_____

_____

_____

9. What is the relationship between the enthalpy of vaporization of a metal and the strength of the bonds that hold the metal together?

_____

_____

_____

10. Draw two diagrams of a metallic bond. In the first diagram, draw a weak metallic bond; in the second, show a metallic bond that would be stronger. Be sure to include nuclear charge and number of electrons in your illustrations.

a.   b.

                    weak bond   **strong bond**

11. Complete the following table:

	Metals	Ionic Compounds
Components		
Overall charge		
Conductive in the solid state		
Melting point		
Hardness		
Malleable		
Ductile		

CHAPTER 6 REVIEW
Chemical Bonding

SECTION 5

SHORT ANSWER Answer the following questions in the space provided.

1. Identify the major assumption of the VSEPR theory, which is used to predict the shape of atoms.

_____

_____

2. In water, two hydrogen atoms are bonded to one oxygen atom. Why isn't water a linear molecule?

_____

_____

3. What orbitals combine together to form sp^3 hybrid orbitals around a carbon atom?

_____

_____

4. What two factors determine whether or not a molecule is polar?

_____

_____

5. Arrange the following types of attractions in order of increasing strength, with 1 being the weakest and 4 the strongest.

_____ hydrogen bonding

_____ ionic

_____ dipole-dipole

_____ London dispersion

6. How are dipole-dipole attractions, London dispersion forces, and hydrogen bonding similar?

_____

_____

_____

_____

Name _____ Date _____ Class _____

7. Complete the following table:

Formula	Lewis structure	Geometry	Polar
H_2S			
CCl_4			
BF_3			
H_2O			
PCl_5			
BeF_2			
SF_6			

CHAPTER 6 REVIEW
Chemical Bonding

MIXED REVIEW

SHORT ANSWER   Answer the following questions in the space provided.

1. Name the type of energy that is a measure of strength for each of the following types of bonds:

_____ **a.** ionic bond

_____ **b.** covalent bond

_____ **c.** metallic bond

2. Use the electronegativity values shown in **Figure 20,** on page 161 of the text, to determine whether each of the following bonds is nonpolar covalent, polar covalent, or ionic.

_____ **a.** H—F   _____ **d.** H—H

_____ **b.** Na—Cl   _____ **e.** H—C

_____ **c.** H—O   _____ **f.** H—N

3. How is a hydrogen bond different from an ionic or covalent bond?

_____

_____

_____

_____

_____

_____

4. H_2S and H_2O have similar structures and their central atoms belong to the same group. Yet H_2S is a gas at room temperature and H_2O is a liquid. Use bonding principles to explain why this is.

_____

_____

_____

_____

MIXED REVIEW continued

5. In what way is a polar-covalent bond similar to an ionic bond?

_____

_____

_____

_____

6. Draw a Lewis structure for each of the following formulas. Determine whether the molecule is polar or nonpolar.

_____ **a.** H_2S

_____ **b.** $COCl_2$

_____ **c.** PCl_3

_____ **d.** CH_2O

MODERN CHEMISTRY

CHAPTER 7 REVIEW

Chemical Formulas and Chemical Compounds

SECTION 1

SHORT ANSWER   Answer the following questions in the space provided.

1. _____ In a Stock system name such as iron(III) sulfate, the Roman numeral tells us

    (a) how many atoms of Fe are in one formula unit.
    (b) how many sulfate ions can be attached to the iron atom.
    (c) the charge on each Fe ion.
    (d) the total positive charge of the formula unit.

2. _____ Changing a subscript in a correctly written chemical formula

    (a) changes the number of moles represented by the formula.
    (b) changes the charges on the other ions in the compound.
    (c) changes the formula so that it no longer represents the compound it previously represented.
    (d) has no effect on the formula.

3. The explosive TNT has the molecular formula $C_7H_5(NO_2)_3$.

_____ **a.** How many elements make up this compound?

_____ **b.** How many oxygen atoms are present in one molecule of $C_7H_5(NO_2)_3$?

_____ **c.** How many atoms in total are present in one molecule of $C_7H_5(NO_2)_3$?

_____ **d.** How many atoms are present in a sample of 2.0×10^{23} molecules of $C_7H_5(NO_2)_3$?

4. How many atoms are present in each of these formula units?

_____ **a.** $Ca(HCO_3)_2$

_____ **b.** $C_{12}H_{22}O_{11}$

_____ **c.** $Fe(ClO_2)_3$

_____ **d.** $Fe(ClO_3)_2$

5. _____ **a.** What is the formula for the compound dinitrogen pentoxide?

_____ **b.** What is the Stock system name for the compound FeO?

_____ **c.** What is the formula for sulfurous acid?

_____ **d.** What is the name for the acid H_3PO_4?

SECTION 1 continued

6. Some binary compounds are ionic, others are covalent. The type of bond favored partially depends on the position of the elements in the periodic table. Label each of these claims as True or False; if False, specify the nature of the error.

   a. Covalently bonded binary molecular compounds are typically composed of nonmetals.

_____

_____

   b. Binary ionic compounds are composed of metals and nonmetals, typically from opposite sides of the periodic table.

_____

_____

7. Refer to **Table 2** on page 226 of the text and **Table 5** on page 230 of the text for examples of names and formulas for polyatomic ions and acids.

   a. Derive a generalization for determining whether an acid name will end in the suffix *-ic* or *-ous*.

_____

_____

_____

   b. Derive a generalization for determining whether an acid name will begin with the prefix *hydro-* or not.

_____

_____

_____

8. Fill in the blanks in the table below.

Compound name	Formula
Aluminum sulfide	
Cesium carbonate	
	$PbCl_2$
	$(NH_4)_3PO_4$
Hydroiodic acid	

CHAPTER 7 REVIEW

Chemical Formulas and Chemical Compounds

SECTION 2

SHORT ANSWER Answer the following questions in the space provided.

1. Assign the oxidation number to the specified element in each of the following examples:

   _____ **a.** S in H_2SO_3

   _____ **b.** S in $MgSO_4$

   _____ **c.** S in K_2S

   _____ **d.** Cu in Cu_2S

   _____ **e.** Cr in Na_2CrO_4

   _____ **f.** N in HNO_3

   _____ **g.** C in $(HCO_3)^-$

   _____ **h.** N in $(NH_4)^+$

2. _____ **a.** What is the formula for the compound sulfur(II) chloride?

   _____ **b.** What is the Stock system name for NO_2?

3. _____ **a.** Use electronegativity values to determine the one element that always has a negative oxidation number when it appears in any binary compound.

   _____ **b.** What is the oxidation number and formula for the element described in part **a** when it exists as a pure element?

4. Tin has possible oxidation numbers of +2 and +4 and forms two known oxides. One of them has the formula SnO_2.

   _____ **a.** Give the Stock system name for SnO_2.

   _____ **b.** Give the formula for the other oxide of tin.

5. Scientists think that two separate reactions contribute to the depletion of the ozone, O_3, layer. The first reaction involves oxides of nitrogen. The second involves free chlorine atoms. The equations that represent the reactions follow. When a compound is not stated as a formula, write the correct formula in the blank beside its name.

   a. _____ (nitrogen monoxide) + $O_3 \rightarrow$ _____ (nitrogen dioxide) + O_2

SECTION 2 continued

    b. $Cl + O_3 \rightarrow$ _____ (chlorine monoxide) $+ O_2$

6. Consider the covalent compound dinitrogen trioxide when answering the following:

_____ **a.** What is the formula for dinitrogen trioxide?

_____ **b.** What is the oxidation number assigned to each nitrogen atom in this compound? Explain your answer.

_____

_____

_____

_____ **c.** Give the Stock name for dinitrogen trioxide.

7. The oxidation numbers assigned to the atoms in some organic compounds have unexpected values. Assign oxidation numbers to each atom in the following compounds: (Note: Some oxidation numbers may not be whole numbers.)

    a. CO_2

_____

    b. CH_4 (methane)

_____

    c. $C_6H_{12}O_6$ (glucose)

_____

    d. C_3H_8 (propane gas)

_____

8. Assign oxidation numbers to each element in the compounds found in the following situations:

    a. Rust, Fe_2O_3, forms on an old nail.

_____

    b. Nitrogen dioxide, NO_2, pollutes the air as a component of smog.

_____

    c. Chromium dioxide, CrO_2, is used to make recording tapes.

_____

CHAPTER 7 REVIEW

Chemical Formulas and Chemical Compounds

SECTION 3

SHORT ANSWER   Answer the following questions in the space provided.

1. Label each of the following statements as True or False:

_____ **a.** If the formula mass of one molecule is x amu, the molar mass is x g/mol.

_____ **b.** Samples of equal numbers of moles of two different chemicals must have equal masses as well.

_____ **c.** Samples of equal numbers of moles of two different molecular compounds must have equal numbers of molecules as well.

2. How many moles of each element are present in a 10.0 mol sample of $Ca(NO_3)_2$?

_____

PROBLEMS   Write the answer on the line to the left. Show all your work in the space provided.

3. Consider a sample of 10.0 g of the gaseous hydrocarbon C_3H_4 to answer the following questions.

_____ **a.** How many moles are present in this sample?

_____ **b.** How many molecules are present in the C_3H_4 sample?

_____ **c.** How many carbon atoms are present in this sample?

SECTION 3 continued

_____ **d.** What is the percentage composition of hydrogen in the sample?

4. One source of aluminum metal is alumina, Al_2O_3.

_____ **a.** Determine the percentage composition of Al in alumina.

_____ **b.** How many pounds of aluminum can be extracted from 2.0 tons of alumina?

5. Compound A has a molar mass of 20 g/mol, and compound B has a molar mass of 30 g/mol.

_____ **a.** What is the mass of 1.0 mol of compound A, in grams?

_____ **b.** How many moles are present in 5.0 g of compound B?

_____ **c.** How many moles of compound B are needed to have the same mass as 6.0 mol of compound A?

CHAPTER 7 REVIEW
Chemical Formulas and Chemical Compounds

SECTION 4

SHORT ANSWER   Answer the following questions in the space provided.

1. Write empirical formulas to match the following molecular formulas:

_____ **a.** $C_2H_6O_4$

_____ **b.** N_2O_5

_____ **c.** Hg_2Cl_2

_____ **d.** C_6H_{12}

2. _____ A certain hydrocarbon has an empirical formula of CH_2 and a molar mass of 56.12 g/mol. What is its molecular formula?

3. A certain ionic compound is found to contain 0.012 mol of sodium, 0.012 mol of sulfur, and 0.018 mol of oxygen.

_____ **a.** What is its empirical formula?

_____ **b.** Is this compound a sulfate, sulfite, or neither?

PROBLEMS   Write the answer on the line to the left. Show all your work in the space provided.

4. Water of hydration was discussed in **Sample Problem K** on pages 243–244 of the text. Strong heating will drive off the water as a vapor in hydrated copper(II) sulfate. Use the data table below to answer the following:

Mass of the empty crucible	4.00 g
Mass of the crucible plus hydrate sample	4.50 g
Mass of the system after heating	4.32 g
Mass of the system after a second heating	4.32 g

_____ **a.** Determine the mass percentage of water in the original sample.

_____ **b.** The compound has the formula $CuSO_4 \cdot xH_2O$. Determine the value of x.

c. What might be the purpose of the second heating?

_____

_____

_____

_____

5. Gas X is found to be 24.0% carbon and 76.0% fluorine by mass.

_____ **a.** Determine the empirical formula of gas X.

_____ **b.** Given that the molar mass of gas X is 200.04 g/mol, determine its molecular formula.

6. A compound is found to contain 43.2% copper, 24.1% chlorine, and 32.7% oxygen by mass.

_____ **a.** Determine its empirical formula.

b. What is the correct Stock system name of the compound in part **a**?

_____

CHAPTER 7 REVIEW

Chemical Formulas and Chemical Compounds

MIXED REVIEW

SHORT ANSWER   Answer the following questions in the space provided.

1. Write formulas for the following compounds:

_____ **a.** copper(II) carbonate

_____ **b.** sodium sulfite

_____ **c.** ammonium phosphate

_____ **d.** tin(IV) sulfide

_____ **e.** nitrous acid

2. Write the Stock system names for the following compounds:

_____ **a.** $Mg(ClO_4)_2$

_____ **b.** $Fe(NO_3)_2$

_____ **c.** $Fe(NO_2)_3$

_____ **d.** CoO

_____ **e.** dinitrogen pentoxide

3. _____ **a.** How many atoms are represented by the formula $Ca(HSO_4)_2$?

_____ **b.** How many moles of oxygen atoms are in a 0.50 mol sample of this compound?

_____ **c.** Assign the oxidation number to sulfur in the HSO_4^- anion.

4. Assign the oxidation number to the element specified in each of the following:

_____ **a.** hydrogen in H_2O_2

_____ **b.** hydrogen in MgH_2

_____ **c.** sulfur in S_8

_____ **d.** carbon in $(CO_3)^{2-}$

_____ **e.** chromium in $Na_2Cr_2O_7$

_____ **f.** nitrogen in NO_2

Name _____ Date _____ Class _____

MIXED REVIEW continued

PROBLEMS Write the answer on the line to the left. Show all your work in the space provided.

5. _____ Following are samples of four different compounds. Arrange them in order of increasing mass, from smallest to largest.

      a. 25 g of oxygen gas   **c.** 3×10^{23} molecules of C_2H_6
      b. 1.00 mol of H_2O **d.** 2×10^{23} molecules of $C_2H_6O_2$

6. _____ **a.** What is the formula for sodium hydroxide?

   _____ **b.** What is the formula mass of sodium hydroxide?

   _____ **c.** What is the mass in grams of 0.25 mol of sodium hydroxide?

7. _____ What is the percentage composition of ethane gas, C_2H_6, to the nearest whole number?

8. _____ Ribose is an important sugar (part of RNA), with a molar mass of 150.15 g/mol. If its empirical formula is CH_2O, what is its molecular formula?

MIXED REVIEW continued

9. Butane gas, C_4H_{10}, is often used as a fuel.

_____ **a.** What is the mass in grams of 3.00 mol of butane?

_____ **b.** How many molecules are present in that 3.00 mol sample?

_____ **c.** What is the empirical formula of the gas?

10. _____ Naphthalene is a soft covalent solid that is often used in mothballs. Its molar mass is 128.18 g/mol and it contains 93.75% carbon and 6.25% hydrogen. Determine the molecular formula of napthalene from this information.

11. Nicotine has the formula $C_xH_yN_z$. To determine its composition, a sample is burned in excess oxygen, producing the following results:

1.0 mol of CO_2
0.70 mol of H_2O
0.20 mol of NO_2

Assume that all the atoms in nicotine are present as products.

_____ **a.** Determine the number of moles of carbon present in the products of this combustion.

Name _____ Date _____ Class _____

MIXED REVIEW continued

_____ **b.** Determine the number of moles of hydrogen present in the combustion products.

_____ **c.** Determine the number of moles of nitrogen present in the combustion products.

_____ **d.** Determine the empirical formula of nicotine based on your calculations.

_____ **e.** In a separate experiment, the molar mass of nicotine is found to be somewhere between 150 and 180 g/mol. Calculate the molar mass of nicotine to the nearest gram.

12. When $MgCO_3(s)$ is strongly heated, it produces solid MgO as gaseous CO_2 is driven off.

_____ **a.** What is the percentage loss in mass as this reaction occurs?

_____ **b.** Assign the oxidation number to each atom in $MgCO_3$.

_____ **c.** Does the oxidation number of carbon change upon the formation of CO_2?

MODERN CHEMISTRY

CHAPTER 8 REVIEW
Chemical Equations and Reactions

SECTION 1

SHORT ANSWER Answer the following questions in the space provided.

1. Match the symbol on the left with its appropriate description on the right.

_____ $\xrightarrow{\Delta}$ **(a)** A precipitate forms.

_____ $\downarrow$ **(b)** A gas forms.

_____ $\uparrow$ **(c)** A reversible reaction occurs.

_____ (l) **(d)** Heat is applied to the reactants.

_____ (aq) **(e)** A chemical is dissolved in water.

_____ $\rightleftarrows$ **(f)** A chemical is in the liquid state.

2. Finish balancing the following equation:

$3Fe_3O_4 +$ _____ $Al \rightarrow$ _____ $Al_2O_3 +$ _____ Fe

3. In each of the following formulas, write the total number of atoms present.

_____ **a.** $4SO_2$

_____ **b.** $8O_2$

_____ **c.** $3Al_2(SO_4)_3$

_____ **d.** $6 \times 10^{23} HNO_3$

4. Convert the following word equation into a balanced chemical equation:
aluminum metal + copper(II) fluoride → aluminum fluoride + copper metal

_____

5. One way to test the salinity of a water sample is to add a few drops of silver nitrate solution with a known concentration. As the solutions of sodium chloride and silver nitrate mix, a precipitate of silver chloride forms, and sodium nitrate is left in solution. Translate these sentences into a balanced chemical equation.

_____

6. a. Balance the following equation: $NaHCO_3(s) \xrightarrow{\Delta} Na_2CO_3(s) + H_2O(g) + CO_2(g)$

_____

SECTION 1 continued

b. Translate the chemical equation in part **a** into a sentence.

_____

_____

_____

_____

7. The poisonous gas hydrogen sulfide, H_2S, can be neutralized with a base such as sodium hydroxide, NaOH. The unbalanced equation for this reaction follows:

$$NaOH(aq) + H_2S(g) \rightarrow Na_2S(aq) + H_2O(l)$$

A student who was asked to balance this equation wrote the following:

$$Na_2OH(aq) + H_2S(g) \rightarrow Na_2S(aq) + H_3O(l)$$

Is this equation balanced? Is it correct? Explain why or why not, and supply the correct balanced equation if necessary.

_____

_____

_____

_____

PROBLEM Write the answer on the line to the left. Show all your work in the space provided.

8. Recall that coefficients in a balanced chemical equation give relative amounts of moles as well as numbers of molecules.

_____ **a.** Calculate the number of moles of CO_2 that form if 10 mol of C_3H_4 react according to the following balanced equation:

$$C_3H_4 + 4O_2 \rightarrow 3CO_2 + 2H_2O$$

_____ **b.** Calculate the number of moles of O_2 that are consumed.

MODERN CHEMISTRY

CHAPTER 8 REVIEW
Chemical Equations and Reactions

SECTION 2

SHORT ANSWER Answer the following questions in the space provided.

1. Match the equation type on the left to its representation on the right.

_____ synthesis

_____ decomposition

_____ single-displacement

_____ double-displacement

(a) $AX + BY \rightarrow AY + BX$

(b) $A + BX \rightarrow AX + B$

(c) $A + B \rightarrow AX$

(d) $AX \rightarrow A + X$

2. _____ In the reaction described by the equation $2Al(s) + 3Fe(NO_3)_2(aq) \rightarrow 3Fe(s) + 2Al(NO_3)_3(aq)$, iron has been replaced by

(a) nitrate.   **(c)** aluminum.
(b) water.   **(d)** nitrogen.

3. _____ Of the following chemical equations, the only reaction that is both synthesis and combustion is

(a) $C(s) + O_2(g) \rightarrow CO_2(g)$.
(b) $2C_4H_{10}(l) + 13O_2(g) \rightarrow 8CO_2(g) + 10H_2O(l)$.
(c) $6CO_2(g) + 6H_2O(g) \rightarrow C_6H_{12}O_6(aq) + 6O_2(g)$.
(d) $C_6H_{12}O_6(aq) + 6O_2(g) \rightarrow 6CO_2(aq) + 6H_2O(l)$.

4. _____ Of the following chemical equations, the only reaction that is both combustion and decomposition is

(a) $S(s) + O_2(g) \rightarrow SO_2(g)$.
(b) $2C_4H_{10}(l) + 13O_2(g) \rightarrow 8CO_2(g) + 10H_2O(l)$.
(c) $2H_2O_2(l) \rightarrow 2H_2O(l) + O_2(g)$.
(d) $2HgO(s) \xrightarrow{\Delta} 2Hg(l) + O_2(g)$.

5. Identify the products when the following substances decompose:

_____ **a.** a binary compound

_____ **b.** most metal hydroxides

_____ **c.** a metal carbonate

_____ **d.** the acid H_2SO_3

6. The complete combustion of a hydrocarbon in excess oxygen yields the products _____ and _____.

SECTION 2 continued

7. For the following four reactions, identify the type, predict the products (make sure formulas are correct), and balance the equations:

   a. $Cl_2(aq) + NaI(aq) \rightarrow$

   _____

   _____

   b. $Mg(s) + N_2(g) \rightarrow$

   _____

   _____

   c. $Co(NO_3)_2(aq) + H_2S(aq) \rightarrow$

   _____

   _____

   d. $C_2H_5OH(aq) + O_2(g) \rightarrow$

   _____

   _____

8. Acetylene gas, C_2H_2, is burned to provide the high temperature needed in welding.

   a. Write the balanced chemical equation for the combustion of C_2H_2 in oxygen.

   _____

   _____

   _____ **b.** If 1.0 mol of C_2H_2 is burned, how many moles of CO_2 are formed?

   _____ **c.** If 1.0 mol of C_2H_2 is burned how many moles of oxygen gas are consumed?

9. **a.** Write the balanced chemical equation for the reaction that occurs when solutions of barium chloride and sodium carbonate are mixed. Refer to **Table 1** on page 437 in **Chapter 13** for solubility.

   _____

   _____

   b. To which of the five basic types of reactions does this reaction belong?

   _____

10. For the commercial preparation of aluminum metal, the metal is extracted by electrolysis from alumina, Al_2O_3. Write the balanced chemical equation for the electrolysis of molten Al_2O_3.

   _____

   _____

Chemical Equations and Reactions

SECTION 3

SHORT ANSWER Answer the following questions in the space provided.

1. List four metals that will *not* replace hydrogen in an acid.

_____

2. Consider the metals iron and silver, both listed in **Table 3** on page 286 of the text. Which one readily forms an oxide in nature, and which one does not?

_____

3. In each of the following pairs, identify the more active element.

_____ **a.** F_2 and I_2

_____ **b.** Mn and K

_____ **c.** Cu and H

4. Use the information in **Table 3** on page 286 of the text to predict whether each of the following reactions will occur. For each reaction that will occur, complete the chemical equation by writing in the products formed and balancing the final equation.

a. $Al(s) + CH_3COOH(aq) \xrightarrow{50°C}$

_____

_____

b. $Al(s) + H_2O(l) \xrightarrow{50°C}$

_____

_____

c. $Cr(s) + CdCl_2(aq) \rightarrow$

_____

_____

d. $Br_2(l) + KCl(aq) \rightarrow$

_____

_____

CHEMICAL EQUATIONS AND REACTIONS **69**

SECTION 3 continued

5. Very active metals will react with water to release hydrogen gas and form hydroxides.

   a. Complete, and then balance, the equation for the reaction of $Ca(s)$ with water.

   _____

   b. The reaction of rubidium, Rb, with water is faster and more violent than the reaction of Na with water. Use the atomic structure and radius of each metal to account for this difference.

   _____

   _____

   _____

   _____

6. Gold, Au, is often used in jewelry. How does the relative activity of Au relate to its use in jewelry?

   _____

   _____

   _____

7. Explain how to use an activity series to predict the outcome of a single-displacement reaction.

   _____

   _____

   _____

   _____

8. Aluminum is above copper in the activity series. Will aluminum metal react with copper(II) nitrate, $Cu(NO_3)_2$, to form aluminum nitrate, $Al(NO_3)_3$? If so, write the balanced chemical equation for the reaction.

   _____

   _____

   _____

   _____

CHAPTER 8 REVIEW

Chemical Equations and Reactions

MIXED REVIEW

SHORT ANSWER Answer the following questions in the space provided.

1. _____ A balanced chemical equation represents all the following *except*

    (a) experimentally established facts.
    (b) the mechanism by which reactants combine to form products.
    (c) identities of reactants and products in a chemical reaction.
    (d) relative quantities of reactants and products in a chemical reaction.

2. _____ According to the law of conservation of mass, the total mass of the reacting substances is

    (a) always more than the total mass of the products.
    (b) always less than the total mass of the products.
    (c) sometimes more and sometimes less than the total mass of the products.
    (d) always equal to the total mass of the products.

3. Predict whether each of the following chemical reactions will occur. For each reaction that will occur, identify the reaction type and complete the chemical equation by writing in the products formed and balancing the final equation. General solubility rules are in **Table 1** on page 437 of the text.

    a. $Ba(NO_3)_2(aq) + Na_3PO_4(aq) \rightarrow$

_____

_____

    b. $Al(s) + O_2(g) \rightarrow$

_____

_____

    c. $I_2(s) + NaBr(aq) \rightarrow$

_____

_____

    d. $C_3H_4(g) + O_2(g) \rightarrow$

_____

_____

   e. electrolysis of molten potassium chloride

_____

_____

4. Some small rockets are powered by the reaction represented by the following unbalanced equation:

$$(CH_3)_2N_2H_2(l) + N_2O_4(g) \rightarrow N_2(g) + H_2O(g) + CO_2(g) + heat$$

  a. Translate this chemical equation into a sentence. (Hint: The name for $(CH_3)_2N_2H_2$ is dimethylhydrazine.)

_____

_____

_____

  b. Balance the formula equation.

_____

_____

5. In the laboratory, you are given two small chips of each of the unknown metals X, Y, and Z, along with dropper bottles containing solutions of $XCl_2(aq)$ and $ZCl_2(aq)$. Describe an experimental strategy you could use to determine the relative activities of X, Y, and Z.

_____

_____

_____

_____

_____

_____

_____

_____

_____

6. List the observations that would indicate that a reaction had occurred.

_____

_____

CHAPTER 9 REVIEW

Stoichiometry

SECTION 1

SHORT ANSWER Answer the following questions in the space provided.

1. _____ The coefficients in a chemical equation represent the

   (a) masses in grams of all reactants and products.
   (b) relative number of moles of reactants and products.
   (c) number of atoms of each element in each compound in a reaction.
   (d) number of valence electrons involved in a reaction.

2. _____ Which of the following would not be studied within the topic of stoichiometry?

   (a) the mole ratio of Al to Cl in the compound aluminum chloride
   (b) the mass of carbon produced when a known mass of sucrose decomposes
   (c) the number of moles of hydrogen that will react with a known quantity of oxygen
   (d) the amount of energy required to break the ionic bonds in CaF_2

3. _____ A balanced chemical equation allows you to determine the

   (a) mole ratio of any two substances in the reaction.
   (b) energy released in the reaction.
   (c) electron configuration of all elements in the reaction.
   (d) reaction mechanism involved in the reaction.

4. _____ The relative number of moles of hydrogen to moles of oxygen that react to form water represents a(n)

   (a) reaction sequence.
   (b) bond energy.
   (c) mole ratio.
   (d) element proportion.

5. Given the reaction represented by the following unbalanced equation: $N_2O(g) + O_2(g) \rightarrow NO_2(g)$

   a. Balance the equation.

   _____

   _____ b. What is the mole ratio of NO_2 to O_2?

   _____ c. If 20.0 mol of NO_2 form, how many moles of O_2 must have been consumed?

   _____ d. Twice as many moles of NO_2 form as moles of N_2O are consumed. True or False?

   _____ e. Twice as many grams of NO_2 form as grams of N_2O are consumed. True or False?

PROBLEMS   Write the answer on the line to the left. Show all your work in the space provided.

6. Given the following equation: $N_2(g) + 3H_2(g) \rightarrow 2NH_3(g)$

_____ **a.** Determine to one decimal place the molar mass of each substance and express each mass in grams per mole.

_____

_____

b. There are six different mole ratios in this system. Write out each one.

_____

_____

7. Given the following equation: $4NH_3(g) + 6NO(g) \rightarrow 5N_2(g) + 6H_2O(g)$

_____ **a.** What is the mole ratio of NO to H_2O?

_____ **b.** What is the mole ratio of NO to NH_3?

_____ **c.** If 0.240 mol of NH_3 react according to the above equation, how many moles of NO will be consumed?

8. Propyne gas can be used as a fuel. The combustion reaction of propyne can be represented by the following equation:

$$C_3H_4(g) + 4O_2(g) \rightarrow 3CO_2(g) + 2H_2O(g)$$

a. Write all the possible mole ratios in this system.

_____

_____

_____

_____

b. Suppose that x moles of water form in the above reaction. The other three mole quantities (*not* in order) are $2x$, $1.5x$, and $0.5x$. Match these quantities to their respective components in the equation above.

_____

_____

CHAPTER 9 REVIEW
Stoichiometry

SECTION 2

PROBLEMS   Write the answer on the line to the left. Show all your work in the space provided.

1. _____ The following equation represents a laboratory preparation for oxygen gas:

$$2KClO_3(s) \rightarrow 2KCl(s) + 3O_2(g)$$

How many moles of O_2 form if 3.0 mol of $KClO_3$ are totally consumed?

2. _____ Given the following equation: $H_2(g) + F_2(g) \rightarrow 2HF(g)$
How many grams of HF gas are produced as 5 mol of fluorine react?

3. _____ Water can be made to decompose into its elements by using electricity according to the following equation:

$$2H_2O(l) \rightarrow 2H_2(g) + O_2(g)$$

How many grams of O_2 are produced when 0.033 mol of water decompose?

4. _____ Sodium metal reacts with water to produce NaOH according to the following equation:

$$2Na(s) + 2H_2O(l) \rightarrow 2NaOH(aq) + H_2(g)$$

How many grams of NaOH are produced if 20.0 g of sodium metal react with excess oxygen?

SECTION 2 continued

5. _____ **a.** What mass of oxygen gas is produced if 100. g of lithium perchlorate are heated and allowed to decompose according to the following equation?

$$LiClO_4(s) \rightarrow LiCl(s) + 2O_2(g)$$

_____ **b.** The oxygen gas produced in part **a** has a density of 1.43 g/L. Calculate the volume of this gas.

6. A car air bag requires 70. L of nitrogen gas to inflate properly. The following equation represents the production of nitrogen gas:

$$2NaN_3(s) \rightarrow 2Na(s) + 3N_2(g)$$

_____ **a.** The density of nitrogen gas is typically 1.16 g/L at room temperature. Calculate the number of grams of N_2 that are needed to inflate the air bag.

_____ **b.** Calculate the number of moles of N_2 that are needed.

_____ **c.** Calculate the number of grams of NaN_3 that must be used to generate the amount of N_2 necessary to properly inflate the air bag.

CHAPTER 9 REVIEW
Stoichiometry

SECTION 3

PROBLEMS   Write the answer on the line to the left. Show all your work in the space provided.

1. _____ The actual yield of a reaction is 22 g and the theoretical yield is 25 g. Calculate the percentage yield.

2. 6.0 mol of N_2 are mixed with 12.0 mol of H_2 according to the following equation:

$$N_2(g) + 3H_2(g) \rightarrow 2NH_3(g)$$

_____ **a.** Which chemical is in excess? What is the excess in moles?

_____ **b.** Theoretically, how many moles of NH_3 will be produced?

_____ **c.** If the percentage yield of NH_3 is 80%, how many moles of NH_3 are actually produced?

3. 0.050 mol of $Ca(OH)_2$ are combined with 0.080 mol of HCl according to the following equation:

$$Ca(OH)_2(aq) + 2HCl(aq) \rightarrow CaCl_2(aq) + 2H_2O(l)$$

_____ **a.** How many moles of HCl are required to neutralize all 0.050 mol of $Ca(OH)_2$?

_____ **b.** What is the limiting reactant in this neutralization reaction?

_____ **c.** How many grams of water will form in this reaction?

4. Acid rain can form in a two-step process, producing $HNO_3(aq)$.

$$N_2(g) + 2O_2(g) \rightarrow 2NO_2(g)$$

$$3NO_2(g) + H_2O(g) \rightarrow 2HNO_3(aq) + NO(g)$$

_____ **a.** A car burns 420. g of N_2 according to the above equations. How many grams of HNO_3 will be produced?

_____ **b.** For the above reactions to occur, O_2 must be in excess in the first step. What is the minimum amount of O_2 needed in grams?

_____ **c.** What volume does the amount of O_2 in part **b** occupy if its density is 1.4 g/L?

CHAPTER 9 REVIEW

Stoichiometry

SHORT ANSWER Answer the following questions in the space provided.

1. Given the following equation: $C_3H_4(g) + xO_2(g) \rightarrow 3CO_2(g) + 2H_2O(g)$

_____ **a.** What is the value of the coefficient x in this equation?

_____ **b.** What is the molar mass of C_3H_4?

_____ **c.** What is the mole ratio of O_2 to H_2O in the above equation?

_____ **d.** How many moles are in an 8.0 g sample of C_3H_4?

_____ **e.** If z mol of C_3H_4 react, how many moles of CO_2 are produced, in terms of z?

2. a. What is meant by *ideal conditions* relative to stoichiometric calculations?

_____

_____

b. What function do ideal stoichiometric calculations serve?

_____

_____

c. Are actual yields typically larger or smaller than theoretical yields?

_____

PROBLEMS Write the answer on the line to the left. Show all your work in the space provided.

3. Assume the reaction represented by the following equation goes all the way to completion:

$$N_2 + 3H_2 \rightarrow 2NH_3$$

_____ **a.** If 6 mol of H_2 are consumed, how many moles of NH_3 are produced?

_____ **b.** How many grams are in a sample of NH_3 that contains 3.0×10^{23} molecules?

c. If 0.1 mol of N_2 combine with H_2, what must be true about the quantity of H_2 for N_2 to be the limiting reactant?

_____

_____

4. _____ If a reaction's theoretical yield is 8.0 g and the actual yield is 6.0 g, what is the percentage yield?

5. Joseph Priestley generated oxygen gas by strongly heating mercury(II) oxide according to the following equation:

$$2HgO(s) \rightarrow 2Hg(l) + O_2(g)$$

_____ **a.** If 15.0 g HgO decompose, how many moles of HgO does this represent?

_____ **b.** How many moles of O_2 are theoretically produced?

_____ **c.** How many grams of O_2 is this?

_____ **d.** If the density of O_2 gas is 1.41 g/L, how many liters of O_2 are produced?

_____ **e.** If the percentage yield is 95.0%, how many grams of O_2 are actually collected?

CHAPTER 10 REVIEW
States of Matter

SECTION 1

SHORT ANSWER   **Answer the following questions in the space provided.**

1. Identify whether the descriptions below describe an ideal gas or a real gas.

_____ **a.** The gas will not condense because the molecules do not attract each other.

_____ **b.** Collisions between molecules are perfectly elastic.

_____ **c.** Gas particles passing close to one another exert an attraction on each other.

2. The formula for kinetic energy is $KE = \frac{1}{2}mv^2$.

a. As long as temperature is constant, what happens to the kinetic energy of the colliding particles during an elastic collision?

_____

b. If two gases have the same temperature and share the same energy but have different molecular masses, which molecules will have the greater speed?

_____

3. Use the kinetic-molecular theory to explain each of the following phenomena:

a. A strong-smelling gas released from a container in the middle of a room is soon detected in all areas of that room.

_____

b. As a gas is heated, its rate of effusion through a small hole increases if all other factors remain constant.

_____

_____

4. a. _____ List the following gases in order of rate of effusion, from lowest to highest. (Assume all gases are at the same temperature and pressure.)

   (a) He   **(b)** Xe   **(c)** HCl   **(d)** Cl_2

SECTION 1 continued

    b. Explain why you put the gases in the order above. Refer to the kinetic-molecular theory to support your explanation.

_____

_____

_____

5. Explain why polar gas molecules experience larger deviations from ideal behavior than nonpolar molecules when all other factors (mass, temperature, etc) are held constant.

_____

_____

_____

6. _____ The two gases in the figure below are simultaneously injected into opposite ends of the tube. The ends are then sealed. They should just begin to mix closest to which labeled point?

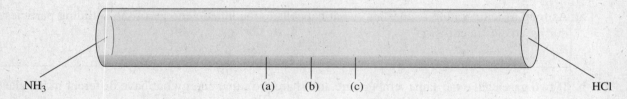

NH_3 (a)   (b)   (c)   HCl

7. Explain the difference in the speed-distribution curves of a gas at the two temperatures shown in the figure below.

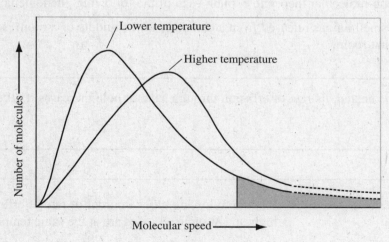

_____

_____

_____

_____

CHAPTER 10 REVIEW
States of Matter

SECTION 2

SHORT ANSWER Answer the following questions in the space provided.

1. _____ Liquids possess all the following properties *except*

    (a) relatively low density.
    (b) the ability to diffuse.
    (c) relative incompressibility.
    (d) the ability to change to a gas.

2. a. Chemists distinguish between intermolecular and intramolecular forces. Explain the difference between these two types of forces.

_____

_____

Classify each of the following as intramolecular or intermolecular:

_____ **b.** hydrogen bonding in liquid water

_____ **c.** the O—H covalent bond in methanol, CH_3OH

_____ **d.** the bonds that cause gaseous Cl_2 to become a liquid when cooled

3. Explain the following properties of liquids by describing what is occurring at the molecular level.

    a. A liquid takes the shape of its container but does not expand to fill its volume.

_____

_____

_____

_____

    b. Polar liquids are slower to evaporate than nonpolar liquids.

_____

_____

_____

SECTION 2 continued

4. Explain briefly why liquids tend to form spherical droplets, decreasing surface area to the smallest size possible.

   _____

   _____

   _____

   _____

5. Is freezing a chemical change or a physical change? Briefly explain your answer.

   _____

   _____

   _____

   _____

6. Is evaporation a chemical or physical change? Briefly explain your answer.

   _____

   _____

   _____

   _____

7. What is the relationship between vaporization and evaporation?

   _____

   _____

   _____

   _____

CHAPTER 10 REVIEW

States of Matter

SECTION 3

SHORT ANSWER Answer the following questions in the space provided.

1. Match description on the right to the correct crystal type on the left.

_____ ionic crystal
    (a) has mobile electrons in the crystal

_____ covalent molecular crystal
    (b) is hard, brittle, and nonconducting

_____ metallic crystal
    (c) typically has the lowest melting point of the four crystal types

_____ covalent network crystal
    (d) has strong covalent bonds between neighboring atoms

2. For each of the four types of solids, give a specific example other than one listed in Table 1 on page 340 of the text.

_____

_____

_____

_____

_____

_____

3. A chunk of solid lead is dropped into a pool of molten lead. The chunk sinks to the bottom of the pool. What does this tell you about the density of the solid lead compared with the density of the molten lead?

_____

4. Answer *amorphous solid* or *crystalline solid* to the following questions:

_____ **a.** Which is less compressible?

_____ **b.** Which has a more clearly defined shape?

_____ **c.** Which is sometimes described as a supercooled liquid?

_____ **d.** Which has a less clearly defined melting point?

SECTION 3 continued

5. Explain the following properties of solids by describing what is occurring at the atomic level.

   a. Metallic solids conduct electricity well, but covalent network solids do not.

   _____

   _____

   _____

   _____

   b. The volume of a solid changes only slightly with a change in temperature or pressure.

   _____

   _____

   _____

   _____

   c. Amorphous solids do not have a definite melting point.

   _____

   _____

   _____

   _____

   d. Ionic crystals are much more brittle than covalent molecular crystals.

   _____

   _____

   _____

   _____

6. Experiments show that it takes 6.0 kJ of energy to melt 1 mol of water ice at its melting point but only about 1.1 kJ to melt 1 mol of methane, CH_4, at its melting point. Explain in terms of intermolecular forces why it takes so much less energy to melt the methane.

   _____

   _____

   _____

   _____

   _____

86 STATES OF MATTER

Name _____ Date _____ Class _____

CHAPTER 10 REVIEW
States of Matter

SECTION 4

SHORT ANSWER Answer the following questions in the space provided.

1. _____ When a substance in a closed system undergoes a phase change and the system reaches equilibrium,

   (a) the two opposing changes occur at equal rates.
   (b) there are no more phase changes.
   (c) one phase change predominates.
   (d) the amount of substance in the two phases changes.

2. Match the following definitions on the right with the words on the left.

   _____ equilibrium   (a) melting

   _____ volatile   (b) opposing changes occurring at equal rates in a closed system

   _____ fusion   (c) readily evaporated

   _____ deposition   (d) a change directly from a gas to a solid

3. Match the process on the right with the change of state on the left.

   _____ solid to gas   (a) melting

   _____ liquid to gas   (b) condensation

   _____ gas to liquid   (c) sublimation

   _____ solid to liquid   (d) vaporization

4. Refer to the phase diagram for water in Figure 16 on page 347 of the text to answer the following questions:

   _____ a. What point represents the conditions under which all three phases can coexist?

   _____ b. What point represents a temperature above which only the vapor phase exists?

   _____ c. Based on the diagram, as the pressure on the water system increases, what happens to the melting point of ice?

   d. What happens when water is at point A on the curve and the temperature increases while the pressure is held constant?

   _____

Name _____ Date _____ Class _____

SECTION 4 continued

5. Use this general equilibrium equation to answer the following questions:

$$\text{reactants} \rightleftarrows \text{products} + \text{energy}$$

_____ **a.** If the forward reaction is favored, will the concentration of reactants increase, decrease, or stay the same?

_____ **b.** If extra product is introduced, which reaction will be favored?

_____ **c.** If the temperature of the system decreases, which reaction will be favored?

6. Refer to the graph below to answer the following questions:

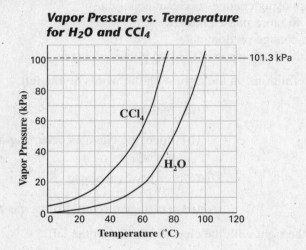

Vapor Pressure vs. Temperature for H_2O and CCl_4

_____ **a.** What is the normal boiling point of CCl_4?

_____ **b.** What would be the boiling point of water if the air pressure over the liquid were reduced to 60 kPa?

_____ **c.** What must the air pressure over CCl_4 be for it to boil at 50°C?

d. Although water has a lower molar mass than CCl_4, it has a lower vapor pressure when measured at the same temperature. What makes water vapor less volatile than CCl_4?

_____

_____

_____

_____

CHAPTER 10 REVIEW
States of Matter

SECTION 5

SHORT ANSWER Answer the following questions in the space provided.

1. Indicate whether each of the following is a *physical* or *chemical* property of water.

_____ **a.** The density of ice is less than the density of liquid water.

_____ **b.** A water molecule contains one atom of oxygen and two atoms of hydrogen.

_____ **c.** There are strong hydrogen bonds between water molecules.

_____ **d.** Ice consists of water molecules in a hexagonal arrangement.

2. Compare a polar water molecule with a less-polar molecule, such as formaldehyde, CH_2O. Both are liquids at room temperature and 1 atm pressure.

_____ **a.** Which liquid should have the higher boiling point?

_____ **b.** Which liquid is more volatile?

_____ **c.** Which liquid has a higher surface tension?

_____ **d.** In which liquid is NaCl, an ionic crystal, likely to be more soluble?

3. Describe hydrogen bonding as it occurs in water in terms of the location of the bond, the particles involved, the strength of the bond, and the effects this type of bonding has on physical properties.

_____

_____

_____

_____

_____

_____

_____

Name _____ Date _____ Class _____

PROBLEMS Write the answer on the line to the left. Show all your work in the space provided.

4. The molar enthalpy of vaporization of water is 40.79 kJ/mol, and the molar enthalpy of fusion of ice is 6.009 kJ/mol. The molar mass of water is 18.02 g/mol.

_____ **a.** How much energy is absorbed when 30.3 g of liquid water boils?

_____ **b.** An energy unit often encountered is the calorie (4.18 J = 1 calorie). Determine the molar enthalpy of fusion of ice in calories per gram.

5. A typical ice cube has a volume of about 16.0 cm³. Calculate the amount of energy needed to melt the ice cube. (Density of ice at 0.°C = 0.917 g/mL; molar enthalpy of fusion of ice = 6.009 kJ/mol; molar mass of H_2O = 18.02 g/mol.)

_____ **a.** Determine the mass of the ice cube.

_____ **b.** Determine the number of moles of H_2O present in the sample.

_____ **c.** Determine the number of kilojoules of energy needed to melt the ice cube.

Name _____ Date _____ Class _____

States of Matter

MIXED REVIEW

SHORT ANSWER   Answer the following questions in the space provided.

1. _____ The average speed of a gas molecule is most directly related to the

    (a) polarity of the molecule.
    (b) pressure of the gas.
    (c) temperature of the gas.
    (d) number of moles in the sample.

2. Use the kinetic-molecular theory to explain the following phenomena:

  a. When 1 mol of a real gas is condensed to a liquid, the volume shrinks by a factor of about 1000.

_____

_____

_____

  b. When a gas in a rigid container is warmed, the pressure on the walls of the container increases.

_____

_____

_____

_____

3. _____ Which of the following statements about liquids and gases is *not* true?

    (a) Molecules in a liquid are much more closely packed than molecules in a gas.
    (b) Molecules in a liquid can vibrate and rotate, but they are bound in fixed positions.
    (c) Liquids are much more difficult to compress into a smaller volume than are gases.
    (d) Liquids diffuse more slowly than gases.

4. Answer *solid* or *liquid* to the following questions:

_____ **a.** Which is less compressible?

_____ **b.** Which is quicker to diffuse into neighboring media?

_____ **c.** Which has a definite volume and shape?

_____ **d.** Which has molecules that are rotating or vibrating primarily in place?

MIXED REVIEW continued

5. Explain why almost all solids are denser than their liquid states by describing what is occurring at the molecular level.

_____

_____

6. A general equilibrium equation for boiling is

$$liquid + energy \rightleftarrows vapor$$

Indicate whether the forward or reverse reaction is favored in each of the following cases:

_____ **a.** The temperature of the system is increased.

_____ **b.** More molecules of the vapor are added to the system.

_____ **c.** The pressure on the system is increased.

7. _____ Freon-11, CCl_3F has been commonly used in air conditioners. It has a molar mass of 137.35 g/mol and its enthalpy of vaporization is 24.8 kJ/mol at its normal boiling point of 24°C. Ideally how much energy in the form of heat is removed from a room by an air conditioner that evaporates 1.00 kg of freon-11?

8. Use the data table below to answer the following:

Composition	Molar mass (g/mol)	Enthalpy vaporization (kJ/mol)	Normal boiling point (°C)	Critical temperature (°C)
He	4	0.08	−269	−268
Ne	20	1.8	−246	−229
Ar	40	6.5	−186	−122
Xe	131	12.6	−107	+17
H_2O	18	40.8	+100	+374
HF	20	25.2	+20	+188
CH_4	16	8.9	−161	−82
C_2H_6	30	15.7	−89	+32

_____ **a.** Among *nonpolar* liquids, those with higher molar masses tend to have normal boiling points that are (higher, lower, or about the same).

_____ **b.** Among compounds of approximately the same molar mass, those with greater polarities tend to have enthalpies of vaporization that are (higher, lower, or about the same).

c. Which is the only noble gas listed that is stable as a liquid at 0°C? Explain your answer using the concept of critical temperature.

_____

_____

CHAPTER 11 REVIEW

Gases

SECTION 1

SHORT ANSWER Answer the following questions in the space provided.

1. _____ $Pressure = \dfrac{force}{surface\ area}$. For a constant force, when the surface area is tripled the pressure is

    (a) doubled.
    (b) a third as much.
    (c) tripled.
    (d) unchanged.

2. _____ Rank the following pressures in increasing order.

    (a) 50 kPa   **(c)** 76 torr
    (b) 2 atm   **(d)** 100 N/m^2

3. Explain how to calculate the partial pressure of a dry gas that is collected over water when the total pressure is atmospheric pressure.

_____

_____

PROBLEMS Write the answer on the line to the left. Show all your work in the space provided.

4. a. Use five to six data points from **Appendix Table A-8** in the text to sketch the curve for water vapor's partial pressure versus temperature on the graph provided below.

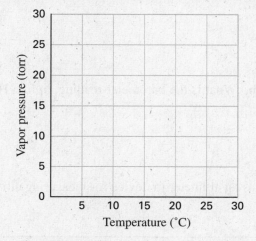

_____ **b.** Do the data points lie on a straight line?

_____ **c.** Based on your sketch, predict the approximate partial pressure for water at 11°C.

SECTION 1 continued

5. Convert a pressure of 0.200 atm to the following units:

_____ **a.** mm Hg

_____ **b.** kPa

6. When an explosive like TNT is detonated, a mixture of gases at high temperature is created. Suppose that gas X has a pressure of 50 atm, gas Y has a pressure of 20 atm, and gas Z has a pressure of 10 atm.

_____ **a.** What is the total pressure in this system?

_____ **b.** What is the total pressure in this system in kPa?

7. The height of the mercury in a barometer is directly proportional to the pressure on the mercury's surface. At sea level, pressure averages 1.0 atm and the level of mercury in the barometer is 760 mm (30. in.). In a hurricane, the barometric reading may fall to as low as 28 in.

_____ **a.** Convert a pressure reading of 28 in. to atmospheres.

_____ **b.** What is the barometer reading, in mm Hg, at a pressure of 0.50 atm?

c. Can a barometer be used as an altimeter (a device for measuring altitude above sea level)? Explain your answer.

_____

_____

_____

MODERN CHEMISTRY

CHAPTER 11 REVIEW

Gases

SECTION 2

SHORT ANSWER   Answer the following questions in the space provided.

1. State whether the pressure of a fixed mass of gas will increase, decrease, or stay the same in the following circumstances:

   _____ **a.** temperature increases, volume stays the same

   _____ **b.** volume increases, temperature stays the same

   _____ **c.** temperature decreases, volume stays the same

   _____ **d.** volume decreases, temperature stays the same

2. Two sealed flasks, A and B, contain two different gases of equal volume at the same temperature and pressure. Assume that flask A is warmed as flask B is cooled. Will the pressure in the two flasks remain equal? If not, which flask will have the higher pressure?

   _____

   _____

   _____

PROBLEMS   Write the answer on the line to the left. Show all your work in the space provided.

3. A bicycle tire is inflated to 55 lb/in.^{2} at 15°C. Assume that the volume of the tire does not change appreciably once it is inflated.

   a. If the tire and the air inside it are heated to 30°C by road friction, does the pressure in the tire increase or decrease? (Assume the volume of air in the tire remains constant.)

   _____

   _____

   b. Because the temperature has doubled, does the pressure double to 110 psi?

   _____

   _____

   c. What will the pressure be when the temperature has doubled? Express your answer in pounds per square inch.

   _____

Name _____ Date _____ Class _____

SECTION 2 continued

4. _____ A 24 L sample of a gas at fixed mass and constant temperature exerts a pressure of 3.0 atm. What pressure will the gas exert if the volume is changed to 16 L?

5. _____ A common laboratory system to study Boyle's law uses a gas trapped in a syringe. The pressure in the system is changed by adding or removing identical weights on the plunger. The original gas volume is 50.0 mL when two weights are present. Predict the new gas volume when four more weights are added.

6. _____ A sample of argon gas occupies a volume of 950 mL at 25.0°C. What volume will the gas occupy at 50.0°C if the pressure remains constant?

7. _____ A 500.0 mL gas sample at STP is compressed to a volume of 300.0 mL and the temperature is increased to 35.0°C. What is the new pressure of the gas in pascals?

8. _____ A sample of gas occupies 1000. mL at standard pressure. What volume will the gas occupy at a pressure of 600. mm Hg if the temperature remains constant?

CHAPTER 11 REVIEW

Gases

_____

SECTION 3

SHORT ANSWER   Answer the following questions in the space provided.

1. _____ The molar mass of a gas at STP is the density of that gas

    (a) multiplied by the mass of 1 mol.   (c) multiplied by 22.4 L.
    (b) divided by the mass of 1 mol.   (d) divided by 22.4 L.

2. _____ For the expression $V = \dfrac{nRT}{P}$, which of the following will cause the volume to increase?

    (a) increasing P (c) increasing T
    (b) decreasing T (d) decreasing n

3. Two sealed flasks, A and B, contain two different gases of equal volume at the same temperature and pressure.

_____ **a.** The two flasks must contain an equal number of molecules. True or False?

_____ **b.** The two samples must have equal masses. True or False?

PROBLEMS   Write the answer on the line to the left. Show all your work in the space provided.

4. Use the data in the table below to answer the following questions.

Formula	Molar mass (g/mol)
N_2	28.02
CO	28.01
C_2H_2	26.04
He	4.00
Ar	39.95

(Assume all gases are at STP.)

_____ **a.** Which gas contains the most molecules in a 5.0 L sample?

_____ **b.** Which gas is the least dense?

_____ **c.** Which two gases have virtually the same density?

_____ **d.** What is the density of N_2 measured at STP?

_____ **a.** How many moles of methane, CH_4 are present in 5.6 L of the gas at STP?

_____ **b.** How many moles of gas are present in 5.6 L of any ideal gas at STP?

_____ **c.** What is the mass of the 5.6 L sample of CH_4?

6. _____ **a.** A large cylinder of He gas, such as that used to inflate balloons, has a volume of 25.0 L at 22°C and 5.6 atm. How many moles of He are in such a cylinder?

_____ **b.** What is the mass of the He calculated in part **a**?

7. When C_3H_4 combusts at STP, 5.6 L of C_3H_4 are consumed according to the following equation:

$$C_3H_4(g) + 4O_2(g) \rightarrow 3CO_2(g) + 2H_2O(l)$$

_____ **a.** How many moles of C_3H_4 react?

_____ **b.** How many moles of O_2, CO_2, and H_2O are either consumed or produced
_____   in the above reaction?
_____

_____ **c.** How many grams of C_3H_4 are consumed?

_____ **d.** How many liters of CO_2 are produced?

_____ **e.** How many grams of H_2O are produced?

Name _____ Date _____ Class _____

CHAPTER 11 REVIEW

Gases

SECTION 4

SHORT ANSWER   Answer the following questions in the space provided.

1. _____ List the following gases in order of increasing rate of effusion. (Assume all gases are at the same temperature and pressure.)

   (a) He   **(b)** Xe   **(c)** HCl   **(d)** Cl_2

2. Explain your reasoning for the order of gases you chose in item **1** above. Refer to the kinetic-molecular theory to support your explanation and cite Graham's law of effusion.

_____

_____

_____

_____

_____

_____

3. _____ The two gases in the figure below are simultaneously injected into opposite ends of the tube. At which labeled point should they just begin to mix?

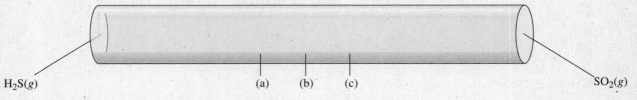

$H_2S(g)$ (a)   (b)   (c) $SO_2(g)$

4. State whether each example describes effusion or diffusion.

_____ **a.** As a puncture occurs, air moves out of a bicycle tire.

_____ **b.** When ammonia is spilled on the floor, the house begins to smell like ammonia.

_____ **c.** The smell of car exhaust pervades an emissions testing station.

5. Describe what happens, in terms of diffusion, when a bottle of perfume is opened.

_____

_____

_____

SECTION 4 continued

PROBLEMS Write the answer on the line to the left. Show all your work in the space provided.

6. _____ **a.** The molar masses of He and of HCl are 4.00 g/mol and 36.46 g/mol, respectively. What is the ratio of the mass of He to the mass of HCl rounded to one decimal place?

_____ **b.** Use your answer in part **a** to calculate the ratio of the average speed of He to the average speed of HCl.

_____ **c.** If helium's average speed is 1200 m/s, what is the average speed of HCl?

7. _____ An unknown gas effuses through an opening at a rate 3.16 times slower than neon gas. Estimate the molar mass of this unknown gas.

CHAPTER 11 REVIEW

Gases

MIXED REVIEW

SHORT ANSWER Answer the following questions in the space provided.

1. Consider the following data table:

Approximate pressure (kPa)	Altitude above sea level (km)
100	0 (sea level)
50	5.5 (peak of Mt. Kilimanjaro)
25	11 (jet cruising altitude)
< 0.1	22 (ozone layer)

  a. Explain briefly why the pressure decreases as the altitude increases.

  _____

  _____

  b. A few places on Earth are below sea level (the Dead Sea, for example). What would be true about the average atmospheric pressure there?

  _____

  _____

  _____

2. Explain how the ideal gas law can be simplified to give Avogadro's law, expressed as $\frac{V}{n} = k$, when the pressure and temperature of a gas are held constant.

  _____

  _____

  _____

PROBLEMS Write the answer on the line to the left. Show all your work in the space provided.

3. Convert a pressure of 0.400 atm to the following units:

  _____ **a.** torr

  _____ **b.** Pa

4. _____ A 250. mL sample of gas is collected at 57°C. What volume will the gas sample occupy at 25°C?

5. _____ H_2 reacts according to the following equation representing the synthesis of ammonia gas:

$$N_2(g) + 3H_2(g) \rightarrow 2NH_3(g)$$

If 1 L of H_2 is consumed, what volume of ammonia will be produced at constant temperature and pressure, based on Gay-Lussac's law of combining volumes?

6. _____ A 7.00 L sample of argon gas at 420. K exerts a pressure of 625 kPa. If the gas is compressed to 1.25 L and the temperature is lowered to 350. K, what will be its new pressure?

7. _____ Chlorine in the upper atmosphere can destroy ozone molecules, O_3. The reaction can be represented by the following equation:

$$Cl_2(g) + 2O_3(g) \rightarrow 2ClO(g) + 2O_2(g)$$

How many liters of ozone can be destroyed at 220. K and 5.0 kPa if 200.0 g of chlorine gas react with it?

8. _____ A gas of unknown molar mass is observed to effuse through a small hole at one-fourth the effusion rate of hydrogen. Estimate the molar mass of this gas. (Round the molar mass of hydrogen to two significant figures.)

MODERN CHEMISTRY

CHAPTER 12 REVIEW
Solutions

SECTION 1

SHORT ANSWER Answer the following questions in the space provided.

1. Match the type of mixture on the left to its representative particle diameter on the right.

_____ solutions   **(a)** larger than 1000 nm

_____ suspensions   **(b)** 1 nm to 1000 nm

_____ colloids   **(c)** smaller than 1 nm

2. Identify the solvent in each of the following examples:

_____ **a.** tincture of iodine (iodine dissolved in ethyl alcohol)

_____ **b.** sea water

_____ **c.** water-absorbing super gels

3. A certain mixture has the following properties:

- No solid settles out during a 48-hour period.
- The path of a flashlight beam is easily seen through the mixture.
- It appears to be homogeneous under a hand lens but not under a microscope.

Is the mixture a suspension, colloid, or true solution? Explain your answer.

_____

_____

_____

_____

4. Define each of the following terms:

a. alloy

_____

_____

b. electrolyte

_____

_____

SECTION 1 continued

    c. aerosol

_____

_____

    d. aqueous solution

_____

_____

5. For each of the following types of solutions, give an example other than those listed in
Table 1 on page 402 of the text:

    a. a gas in a liquid

_____

_____

    b. a liquid in a liquid

_____

_____

    c. a solid in a liquid

_____

_____

6. Using the following models of solutions shown at the particle level, indicate which will conduct
electricity. Give a reason for each model.

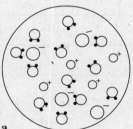
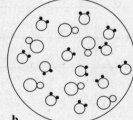
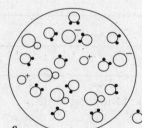

    a.   **b.**   **c.**   ⊙ = water molecule

    a. _____

    b. _____

    c. _____

CHAPTER 12 REVIEW
Solutions

SECTION 2

SHORT ANSWER Answer the following questions in the space provided.

1. The following are statements about the dissolving process. Explain each one at the molecular level.

   a. Increasing the pressure of a solute gas above a liquid solution increases the solubility of the gas in the liquid.

   _____

   _____

   _____

   b. Increasing the temperature of water speeds up the rate at which many solids dissolve in this solvent.

   _____

   _____

   _____

   _____

   _____

   c. Increasing the surface area of a solid solute speeds up the rate at which it dissolves in a liquid solvent.

   _____

   _____

   _____

   _____

2. The solubility of $KClO_3$ at 25°C is 10. g of solute per 100. g of H_2O.

   a. If 15 g of $KClO_3$ are stirred into 100 g of water at 25°C, how much of the $KClO_3$ will dissolve? Is the solution saturated, unsaturated, or supersaturated?

   _____

   _____

SECTION 2 continued

   b. If 15 g of $KClO_3$ are stirred into 200 g of water at 25°C, how much of the $KClO_3$ will dissolve? Is the solution saturated, unsaturated, or supersaturated?

_____

_____

PROBLEMS Write the answer on the line to the left. Show all your work in the space provided.

3. Use the data in **Table 4** on page 410 of the text to answer the following questions:

_____ **a.** How many grams of LiCl are needed to make a saturated solution with 300. g of water at 20°C?

_____ **b.** What is the minimum amount of water needed to dissolve 51 g of $NaNO_3$ at 40°C?

_____ **c.** Which solute forms a saturated solution when 36 g of it are dissolved in 25 g of water at 20°C?

4. KOH is an ionic solid readily soluble in water.

_____ **a.** What is its enthalpy of solution in kJ/g? Refer to the data in **Table 5** on page 416 of the text.

   b. Will the temperature of the system increase or decrease as the dissolution of KOH proceeds? Why?

_____

_____

_____

CHAPTER 12 REVIEW
Solutions

SHORT ANSWER   Answer the following questions in the space provided.

1. Describe the errors made by the following students in making molar solutions.

   a. James needs a 0.600 M solution of KCl. He measures out 0.600 g of KCl and adds 1 L of water to the solid.

   _____

   _____

   _____

   _____

   b. Mary needs a 0.02 M solution of $NaNO_3$. She calculates that she needs 2.00 g of $NaNO_3$ for 0.02 mol. She puts this solid into a 1.00 L volumetric flask and fills the flask to the 1.00 L mark.

   _____

   _____

   _____

   _____

PROBLEMS   Write the answer on the line to the left. Show all of your work in the space provided.

2. _____   What is the molarity of a solution made by dissolving 2.0 mol of solute in 6.0 L of solvent?

3. _____ CH_3OH is soluble in water. What is the molality of a solution made by dissolving 8.0 g of CH_3OH in 250. g of water?

SECTION 3 continued

4. Marble chips effervesce when treated with hydrochloric acid. This reaction is represented by the following equation:

$$CaCO_3(s) + 2HCl(aq) \rightarrow CaCl_2(aq) + CO_2(g) + H_2O(l)$$

To produce a reaction, 25.0 mL of 4.0 M HCl is added to excess $CaCO_3$.

_____ **a.** How many moles of HCl are consumed in this reaction?

_____ **b.** How many liters of CO_2 are produced at STP?

_____ **c.** How many grams of $CaCO_3$ are consumed?

5. Tincture of iodine is $I_2(s)$ dissolved in ethanol, C_2H_5OH. A 1% solution of tincture of iodine is 10.0 g of solute for 1000. g of solution.

_____ **a.** How many grams of solvent are present in 1000. g of this solution?

_____ **b.** How many moles of solute are in 10.0 g of I_2?

_____ **c.** What is the molality of this 1% solution?

d. To determine a solution's molarity, the density of that solution can be used. Explain how you would use the density of the tincture of iodine solution to calculate its molarity.

_____

_____

_____

_____

CHAPTER 12 REVIEW
Solutions

MIXED REVIEW

SHORT ANSWER   Answer the following questions in the space provided.

1. Solid $CaCl_2$ does not conduct electricity. Explain why it is considered to be an electrolyte.

_____

_____

_____

2. Explain the following statements at the molecular level:

a. Generally, a polar liquid and a nonpolar liquid are immiscible.

_____

_____

_____

b. Carbonated soft drinks taste flat when they warm up.

_____

_____

_____

_____

3. An unknown compound is observed to mix with toluene, $C_6H_5CH_3$, but not with water.

a. Is the unknown compound ionic, polar covalent, or nonpolar covalent? Explain your answer.

_____

_____

b. Suppose the unknown compound is also a liquid. Will it be able to dissolve table salt? Explain why or why not.

_____

_____

_____

MIXED REVIEW continued

PROBLEMS   Write the answer on the line to the left. Show all your work in the space provided.

4. Consider 500. mL of a 0.30 M $CuSO_4$ solution.

_____ **a.** How many moles of solute are present in this solution?

_____ **b.** How many grams of solute were used to prepare this solution?

5. a. If a solution is electrically neutral, can all of its ions have the same type of charge? Explain your answer.

_____

_____

_____ **b.** The concentration of the OH^- ions in pure water is known to be 1.0×10^{-7} M. How many OH^- ions are present in each milliliter of pure water?

6. 90. g of $CaBr_2$ are dissolved in 900. g of water.

_____ **a.** What volume does the 900. g of water occupy if its density is 1.00 g/mL?

_____ **b.** What is the molality of this solution?

CHAPTER 13 REVIEW

Ions in Aqueous Solutions and Colligative Properties

SECTION 1

SHORT ANSWER Answer the following questions in the space provided.

1. Use the guidelines in **Table 1** on page 437 of the text to predict the solubility of the following compounds in water:

_____ **a.** magnesium nitrate

_____ **b.** barium sulfate

_____ **c.** calcium carbonate

_____ **d.** ammonium phosphate

2. 1.0 mol of magnesium acetate is dissolved in water.

_____ **a.** Write the formula for magnesium acetate.

_____ **b.** How many moles of ions are released into solution?

_____ **c.** How many moles of ions are released into a solution made from 0.20 mol magnesium acetate dissolved in water?

3. Write the formula for the precipitate formed

_____ **a.** when solutions of magnesium chloride and potassium phosphate are combined.

_____ **b.** when solutions of sodium sulfide and silver nitrate are combined.

4. Write ionic equations for the dissolution of the following compounds:

a. $Na_3PO_4(s)$

_____

b. iron(III) sulfate(s)

_____

5. a. Write the net ionic equation for the reaction that occurs when solutions of lead(II) nitrate and ammonium sulfate are combined.

_____

b. What are the spectator ions in this system?

_____

SECTION 1 continued

6. The following solutions are combined in a beaker: $NaCl$, Na_3PO_4, and $Ba(NO_3)_2$.

   a. Will a precipitate form when the above solutions are combined? If so, write the name and formula of the precipitate.

   _____

   _____

   b. List all spectator ions present in this system.

   _____

   _____

7. It is possible to have spectator ions present in many chemical systems, not just in precipitation reactions. Consider this example:

$$Al(s) + HCl(aq) \rightarrow AlCl_3(aq) + H_2(g) \text{ (unbalanced)}$$

   _____ a. In an aqueous solution of HCl, virtually every HCl molecule is ionized. True or False?

   _____ b. There is only one spectator ion in this system. Is it $Al^{3+}(aq)$, $H^+(aq)$, or $Cl^-(aq)$?

   c. Balance the above equation.

   _____

   _____ d. If 9.0 g of Al metal react with excess HCl according to the balanced equation in part **c**, what volume of hydrogen gas at STP will be produced? Show all your work.

8. Acetic acid, CH_3CO_2H, is a weak electrolyte. Write an equation to represent its ionization in water. Include the hydronium ion, H_3O^+.

   _____

CHAPTER 13 REVIEW

Ions in Aqueous Solutions and Colligative Properties

SECTION 2

PROBLEMS   Write the answer on the line to the left. Show all your work in the space provided.

1. _____   **a.** Predict the boiling point of a 0.200 *m* solution of glucose in water.

_____   **b.** Predict the boiling point of a 0.200 *m* solution of potassium iodide in water.

2. A chief ingredient of antifreeze is liquid ethylene glycol, $C_2H_4(OH)_2$. Assume $C_2H_4(OH)_2$ is added to a car radiator that is holding 5.0 kg of water.

_____   **a.** How many moles of ethylene glycol should be added to the radiator to lower the freezing point of the water from 0°C to −18°C?

_____   **b.** How many grams of ethylene glycol does the quantity in part **a** represent?

_____   **c.** Ethylene glycol has a density of 1.1 kg/L. How many liters of $C_2H_4(OH)_2$ should be added to the water in the radiator to prevent freezing down to −18°C?

SECTION 2 continued

    d. In World War II, soldiers in the Sahara Desert needed a supply of antifreeze to protect the radiators of their vehicles. The temperature in the Sahara almost never drops to 0°C, so why was the antifreeze necessary?

_____

_____

_____

_____

3. An important use of colligative properties is to determine the molar mass of unknown substances. The following situation is an example: 12.0 g of unknown compound X, a nonpolar nonelectrolyte, is dissolved in 100.0 g of melted camphor. The resulting solution freezes at 99.4°C. Consult **Table 2** on page 448 of the text for any other data needed to answer the following questions:

_____ **a.** By how many °C did the freezing point of camphor change from its normal freezing point?

_____ **b.** What is the molality of the solution of camphor and compound X, based on freezing-point data?

_____ **c.** If there are 12.0 g of compound X per 100.0 g of camphor, how many grams of compound X are there per kilogram of camphor?

_____ **d.** What is the molar mass of compound X?

4. Explain why the ability of a solution to conduct an electric current is not a colligative property.

_____

_____

_____

CHAPTER 13 REVIEW

Ions in Aqueous Solutions and Colligative Properties

MIXED REVIEW

SHORT ANSWER   Answer the following questions in the space provided.

1. Match the four compounds on the right to their descriptions on the left.

   _____ an ionic compound that is quite soluble in water   **(a)** HCl

   _____ an ionic compound that is not very soluble in water   **(b)** $NaNO_3$

   _____ a molecular compound that ionizes in water   **(c)** AgCl

   _____ a molecular compound that does not ionize in water   **(d)** C_2H_5OH

2. Consider nonvolatile nonelectrolytes dissolved in various liquid solvents to complete the following statements:

   _____ **a.** The change in the boiling point does *not* vary with the identity of the _____ (solute, solvent), assuming all other factors remain constant.

   _____ **b.** The change in the boiling point varies with the identity of the _____ (solute, solvent), assuming all other factors remain constant.

   _____ **c.** The change in the boiling point becomes greater as the concentration of the solute in solution _____ (increases, decreases).

3. **a.** Name two compounds in solution that could be combined to cause the formation of a calcium carbonate precipitate.

   _____

   _____

   b. Identify any spectator ions in the system you described in part **a.**

   _____

   c. Write the net ionic equation for the formation of calcium carbonate.

   _____

4. Explain why applying rock salt (impure NaCl) to an icy sidewalk hastens the melting process.

   _____

   _____

MIXED REVIEW continued

PROBLEMS   Write the answer on the line to the left. Show all your work in the space provided.

5. _____   Some insects survive cold winters by generating an antifreeze inside their cells. The antifreeze produced is glycerol, $C_3H_5(OH)_3$, a nonvolatile nonelectrolyte that is quite soluble in water. What must the molality of a glycerol solution be to lower the freezing point of water to $-25.0°C$?

6. _____   How many grams of methanol, CH_3OH, should be added to 200. g of acetic acid to lower its freezing point by $1.30°C$? Refer to **Table 2** on page 448 of the text for any necessary data.

7. _____   The boiling point of a solution of glucose, $C_6H_{12}O_6$, and water was recorded to be $100.34°C$. Calculate the molality of this solution.

8. HF(*aq*) is a weak acid. A 0.05 mol sample of HF is added to 1.0 kg of water.

   a. Write the equation for the ionization of HF to form hydronium ions.

   _____

   _____   **b.** If HF became 100% ionized, how many moles of its ions would be released?

9. _____   Which solution has the highest osmotic pressure?

   a. 0.1 *m* glucose
   b. 0.1 *m* sucrose
   c. 0.5 *m* glucose
   d. 0.2 *m* sucrose

CHAPTER 14 REVIEW

Acids and Bases

SECTION 1

SHORT ANSWER   Answer the following questions in the space provided.

1. Name the following compounds as acids:

_____ **a.** H_2SO_4

_____ **b.** H_2SO_3

_____ **c.** H_2S

_____ **d.** $HClO_4$

_____ **e.** hydrogen cyanide

2. _____ Which (if any) of the acids mentioned in item **1** are binary acids?

3. Write formulas for the following acids:

_____ **a.** nitrous acid

_____ **b.** hydrobromic acid

_____ **c.** phosphoric acid

_____ **d.** acetic acid

_____ **e.** hypochlorous acid

4. Calcium selenate has the formula $CaSeO_4$.

_____ **a.** What is the formula for selenic acid?

_____ **b.** What is the formula for selenous acid?

5. Use an activity series to identify two metals that will not generate hydrogen gas when treated with an acid.

_____

6. Write balanced chemical equations for the following reactions of acids and bases:

a. aluminum metal with dilute nitric acid

_____

b. calcium hydroxide solution with acetic acid

SECTION 1 continued

7. Write net ionic equations that represent the following reactions:

   a. the ionization of $HClO_3$ in water

   _____

   b. NH_3 functioning as an Arrhenius base

   _____

8. a. Explain how strong acid solutions carry an electric current.

   _____

   _____

   _____

   b. Will a strong acid or a weak acid conduct electricity better, assuming all other factors remain constant? Explain why one is a better conductor.

   _____

   _____

   _____

9. Most acids react with solid carbonates, as in the following equation:

$$CaCO_3(s) + HCl(aq) \rightarrow CaCl_2(aq) + H_2O(l) + CO_2(g) \text{ (unbalanced)}$$

   a. Balance the above equation.

   _____

   b. Write the net ionic equation for the above reaction.

   _____

   _____ **c.** Identify all spectator ions in this system.

   _____ **d.** How many liters of CO_2 form at STP if 5.0 g of $CaCO_3$ are treated with excess hydrochloric acid? Show all your work.

CHAPTER 14 REVIEW

Acids and Bases

SECTION 2

SHORT ANSWER Answer the following questions in the space provided.

1. a. Write the two equations that show the two-stage ionization of sulfurous acid in water.

_____

_____

b. Which stage of ionization usually produces more ions? Explain your answer.

_____

_____

_____

2. a. Define a Lewis base. Can OH$^-$ function as a Lewis base? Explain your answer.

_____

_____

_____

_____

b. Define a Lewis acid. Can H$^+$ function as a Lewis acid? Explain your answer.

_____

_____

_____

3. Identify the Brønsted-Lowry acid and the Brønsted-Lowry base on the reactant side of each of the following equations for reactions that occur in aqueous solution. Explain your answers.

a. $H_2O(l) + HNO_3(aq) \rightarrow H_3O^+(aq) + NO_3^-(aq)$

_____

_____

b. $HF(aq) + HS^-(aq) \rightarrow H_2S(aq) + F^-(aq)$

_____

_____

SECTION 2 continued

4. a. Write the equation for the first ionization of H_2CO_3 in aqueous solution. Assume that water serves as the reactant that attaches to the hydrogen ion released from the H_2CO_3. Which of the reactants is the Brønsted-Lowry acid, and which is the Brønsted-Lowry base? Explain your answer.

_____

_____

_____

b. Write the equation for the second ionization, that of the ion that was formed by the H_2CO_3 in the reaction you described above. Again, assume that water serves as the reactant that attaches to the hydrogen ion released. Which of the reactants is the Brønsted-Lowry acid, and which is the Brønsted-Lowry base? Explain your answer.

_____

_____

_____

c. What is the name for a substance, such as H_2CO_3, that can donate two protons?

_____

5. a. How many electron pairs surround an atom of boron (B, element 5) bonded in the compound BCl_3?

_____

b. How many electron pairs surround an atom of nitrogen (N, element 7) in the compound NF_3?

_____

c. Write an equation for the reaction between the two compounds above. Assume that they react in a 1:1 ratio to form one molecule as product.

_____

d. Assuming that the B and the N are covalently bonded to each other in the product, which of the reactants is the Lewis acid? Is this reactant also a Brønsted-Lowry acid? Explain your answers.

_____

_____

e. Which of the reactants is the Lewis base? Explain your answer.

_____

_____

CHAPTER 14 REVIEW

Acids and Bases

SECTION 3

SHORT ANSWER   Answer the following questions in the space provided.

1. Answer the following questions according to the Brønsted-Lowry definitions of acids and bases:

_____ **a.** What is the conjugate base of H_2SO_3?

_____ **b.** What is the conjugate base of NH_4^+?

_____ **c.** What is the conjugate base of H_2O?

_____ **d.** What is the conjugate acid of H_2O?

_____ **e.** What is the conjugate acid of $HAsO_4^{2-}$?

2. Consider the reaction described by the following equation:

$$NH_4^+(aq) + CO_3^{2-}(aq) \rightleftarrows NH_3(aq) + HCO_3^-(aq)$$

a. If NH_4^+ is considered acid 1, identify the other three terms as acid 2, base 1, and base 2 to indicate the conjugate acid-base pairs.

_____ CO_3^{2-}

_____ HCO_3^-

_____ NH_3

_____ **b.** A proton has been transferred from acid 1 to base 2 in the above reaction. True or False?

3. Consider the neutralization reaction described by the equation: $HCO_3^-(aq) + OH^-(aq) \rightleftarrows CO_3^{2-}(aq) + H_2O(l)$

a. Label the conjugate acid-base pairs in this system.

_____

_____

b. Is the forward or reverse reaction favored? Explain your answer.

_____

_____

_____

_____

Name _____ Date _____ Class _____

SECTION 3 continued

4. **Table 6** on page 485 of the text lists several amphoteric species, but only one other than water is neutral.

_____ **a.** Identify that neutral compound.

b. Write two equations that demonstrate this compound's amphoteric properties.

_____

_____

_____

5. Write the formula for the salt formed in each of the following neutralization reactions:

_____ **a.** potassium hydroxide combined with phosphoric acid

_____ **b.** calcium hydroxide combined with nitrous acid

_____ **c.** hydrobromic acid combined with barium hydroxide

_____ **d.** lithium hydroxide combined with sulfuric acid

6. Consider the following unbalanced equation for a neutralization reaction:

$$H_2SO_4(aq) + NaOH(aq) \rightarrow Na_2SO_4(aq) + H_2O(l)$$

a. Balance the equation.

_____

_____ **b.** In this system there are two spectator ions. Identify them.

_____ **c.** For the reaction to completely consume all reactants, what should be the mole ratio of acid to base?

7. The gases that produce acid rain are often referred to as NO_x and SO_x.

a. List three specific examples of these gases.

_____

b. Coal- and oil-burning power plants oxidize any sulfur in their fuel as it burns in air, and this forms SO_2 gas. The SO_2 is further oxidized by O_2 in our atmosphere, forming SO_3 gas. The SO_3 gas can combine with water to form sulfuric acid. Write balanced chemical equations to illustrate these three reactions.

_____

_____

_____

c. Industrial plants making fertilizers and detergents release nitrogen oxide gases into the air. Write a balanced equation for converting $N_2O_5(g)$ into nitric acid by reacting it with water.

_____

122 ACIDS AND BASES

MODERN CHEMISTRY

CHAPTER 14 REVIEW
Acids and Bases

MIXED REVIEW

SHORT ANSWER   Answer the following questions in the space provided.

1. _____ **a.** Write the formula for hypochlorous acid.

_____ **b.** Write the name for HF(*aq*).

_____ **c.** If $Pb(C_2O_4)_2$ is lead(IV) oxalate, what is the formula for oxalic acid?

_____ **d.** Name the acid that is present in vinegar.

2. Answer the following questions according to the Brønsted-Lowry acid-base theory. Consult **Table 6** on page 485 of the text as needed.

_____ **a.** What is the conjugate base of H_2S?

_____ **b.** What is the conjugate base of HPO_4^{2-}?

_____ **c.** What is the conjugate acid of NH_3?

3. Consider the reaction represented by the following equation:

$$OH^-(aq) + HCO_3^-(aq) \rightarrow H_2O(l) + CO_3^{2-}(aq)$$

If OH^- is considered base 1, what are acid 1, acid 2, and base 2?

_____ **a.** acid 1

_____ **b.** acid 2

_____ **c.** base 2

4. Write the formula for the salt that is produced in each of the following neutralization reactions:

_____ **a.** sulfurous acid combined with potassium hydroxide

_____ **b.** calcium hydroxide combined with phosphoric acid

5. Carbonic acid releases H_3O^+ ions into water in two stages.

a. Write equations representing each stage.

_____

_____

_____ **b.** Which stage releases more ions into solution?

MIXED REVIEW continued

6. Glacial acetic acid is a highly viscous liquid that is close to 100% CH_3COOH. When it mixes with water, it forms dilute acetic acid.

   a. When making a dilute acid solution, should you add acid to water or water to acid? Explain your answer.

   _____

   _____

   _____

   b. Glacial acetic acid does not conduct electricity, but dilute acetic acid does. Explain this statement.

   _____

   _____

   _____

   c. Dilute acetic acid does not conduct electricity as well as dilute nitric acid at the same concentration. Is acetic acid a strong or weak acid?

   _____

   d. Although there are four H atoms per molecule, acetic acid is monoprotic. Show the structural formula for CH_3COOH, and indicate the H atom that ionizes.

   e. _____ How many grams of glacial acetic acid should be used to make 250 mL of 2.00 M acetic acid? Show all your work.

7. The overall effect of acid rain on lakes and ponds is partially determined by the geology of the lake bed. In some cases, the rock is limestone, which is rich in calcium carbonate. Calcium carbonate reacts with the acid in lake water according to the following (incomplete) ionic equation:

$$CaCO_3(s) + 2H_3O^+(aq) \rightarrow$$

   a. Complete the ionic equation begun above.

   _____

   b. If this reaction is the only reaction involving H_3O^+ occurring in the lake, does the concentration of H_3O^+ in the lake water increase or decrease? What effect does this have on the acidity of the lake water?

   _____

CHAPTER 15 REVIEW
Acid-Base Titration and pH

SECTION 1

SHORT ANSWER Answer the following questions in the space provided.

1. Calculate the following values without using a calculator.

_____ **a.** The $[H_3O^+]$ is 1×10^{-6} M in a solution. Calculate the $[OH^-]$.

_____ **b.** The $[H_3O^+]$ is 1×10^{-9} M in a solution. Calculate the $[OH^-]$.

_____ **c.** The $[OH^-]$ is 1×10^{-12} M in a solution. Calculate the $[H_3O^+]$.

_____ **d.** The $[OH^-]$ in part **c** is reduced by half, to 0.5×10^{-12} M. Calculate the $[H_3O^+]$.

_____ **e.** The $[H_3O^+]$ and $[OH^-]$ are _____ (directly, inversely, or not) proportional in any system involving water.

2. Calculate the following values without using a calculator.

_____ **a.** The pH of a solution is 2.0. Calculate the pOH.

_____ **b.** The pOH of a solution is 4.73. Calculate the pH.

_____ **c.** The $[H_3O^+]$ in a solution is 1×10^{-3} M. Calculate the pH.

_____ **d.** The pOH of a solution is 5.0. Calculate the $[OH^-]$.

_____ **e.** The pH of a solution is 1.0. Calculate the $[OH^-]$.

3. Calculate the following values.

_____ **a.** The $[H_3O^+]$ is 2.34×10^{-5} M in a solution. Calculate the pH.

_____ **b.** The pOH of a solution is 3.5. Calculate the $[OH^-]$.

_____ **c.** The $[H_3O^+]$ is 4.6×10^{-8} M in a solution. Calculate the $[OH^-]$.

PROBLEMS Write the answer on the line to the left. Show all your work in the space provided.

4. $[H_3O^+]$ in an aqueous solution = 2.3×10^{-3} M.

_____ **a.** Calculate $[OH^-]$ in this solution.

ACID-BASE TITRATION AND pH **125**

_____ **b.** Calculate the pH of this solution.

_____ **c.** Calculate the pOH of this solution.

d. Is the solution acidic, basic, or neutral? Explain your answer.

_____

5. Consider a dilute solution of 0.025 M $Ba(OH)_2$ in answering the following questions.

   a. What is the $[OH^-]$ in this solution? Explain your answer.

   _____

   _____

   _____ **b.** What is the pH of this solution?

6. Vinegar purchased in a store may contain 6 g of CH_3COOH per 100 mL of solution.

   _____ **a.** What is the molarity of the solute?

   b. The actual $[H_3O^+]$ in the vinegar solution in part **a** is 4.2×10^{-3} M. In this solution, has more than 1% or less than 1% of the acetic acid ionized? Explain your answer.

   _____

   _____

   _____ **c.** Is acetic acid strong or weak, based on the ionization information from part **b**?

   _____ **d.** What is the pH of this vinegar solution?

Name _____ Date _____ Class _____

Acid-Base Titration and pH

SECTION 2

SHORT ANSWER   Answer the following questions in the space provided.

1. Below is a pH curve from an acid-base titration. On it are labeled three points: X, Y, and Z.

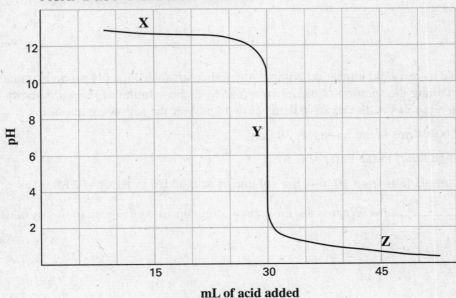

Acid-Base Titration Curve

_____ **a.** Which point represents the equivalence point?

_____ **b.** At which point is there excess acid in the system?

_____ **c.** At which point is there excess base in the system?

_____ **d.** If the base solution is 0.250 M and there is one equivalent of OH^- ions for each mole of base, how many moles of OH^- ions are consumed at the end point of the titration?

PROBLEMS   Write the answer on the line to the left. Show all your work in the space provided.

2. A standardized solution of 0.065 M HCl is titrated with a saturated solution of calcium hydroxide to determine its molarity and its solubility. It takes 25.0 mL of the base to neutralize 10.0 mL of the acid.

a. Write the balanced molecular equation for this neutralization reaction.

_____

_____ **b.** Determine the molarity of the $Ca(OH)_2$ solution.

_____ **c.** Based on your answer to part **b**, calculate the solubility of the base in grams per liter of solution. (Hint: What is the concentration of $Ca(OH)_2$ in the saturated solution?)

3. It is possible to carry out a titration without any indicator. Instead, a pH probe is immersed in a beaker containing the solution of unknown molarity, and a solution of known molarity is slowly added from a buret. Use the titration data below to answer the following questions.

Volume of $KOH(aq)$ in the beaker = 30.0 mL

Molarity of $HCl(aq)$ in the buret = 0.50 M

At the instant pH falls from 10 to 4, the volume of acid added to KOH = 27.8 mL.

_____ **a.** What is the mole ratio of chemical equivalents in this system?

_____ **b.** Calculate the molarity of the KOH solution, based on the above data.

CHAPTER 15 REVIEW
Acid-Base Titration and pH

MIXED REVIEW

SHORT ANSWER Answer the following questions in the space provided.

1. Calculate the following values without using a calculator.

_____ **a.** The $[H_3O^+]$ in a solution is 1×10^{-4} M. Calculate the pH.

_____ **b.** The pH of a solution is 13.0. Calculate the $[H_3O^+]$.

_____ **c.** The $[OH^-]$ in a solution is 1×10^{-5} M. Calculate the $[H_3O^+]$.

_____ **d.** The pH of a solution is 4.72. Calculate the pOH.

_____ **e.** The $[OH^-]$ in a solution is 1.0 M. Calculate the pH.

2. Calculate the following values.

_____ **a.** The $[H_3O^+]$ in a solution is 6.25×10^{-9} M. Calculate the pH.

_____ **b.** The pOH of a solution is 2.34. Calculate the $[OH^-]$.

_____ **c.** The pH of milk of magnesia is approximately 10.5. Calculate the $[OH^-]$.

PROBLEMS Write the answer on the line to the left. Show all your work in the space provided.

3. A 0.0012 M solution of H_2SO_4 is 100% ionized.

_____ **a.** What is the $[H_3O^+]$ in the H_2SO_4 solution?

_____ **b.** What is the $[OH^-]$ in this solution?

_____ **c.** What is the pH of this solution?

4. In a titration, a 25.0 mL sample of 0.150 M HCl is neutralized with 44.45 mL of $Ba(OH)_2$.

   a. Write the balanced molecular equation for this reaction.

   _____

   _____ **b.** What is the molarity of the base solution?

5. 3.09 g of boric acid, H_3BO_3, are dissolved in 200 mL of solution.

   _____ **a.** Calculate the molarity of the solution.

   b. H_3BO_3 ionizes in solution in three stages. Write the equation showing the ionization for each stage. Which stage proceeds furthest to completion?

   _____

   _____

   _____

   _____

   _____ **c.** What is the $[H_3O^+]$ in this boric acid solution if the pH = 4.90?

   _____ **d.** Is the percentage ionization of this H_3BO_3 solution more than or less than 1%?

CHAPTER 16 REVIEW

Reaction Energy

SECTION 1

SHORT ANSWER   Answer the following questions in the space provided.

1. For elements in their standard state, the value of $\triangle H_f^{\,o}$ is _____.

2. The formation and decomposition of water can be represented by the following thermochemical equations:

$$H_2(g) + \tfrac{1}{2}O_2(g) \rightarrow H_2O(g) + 241.8 \text{ kJ/mol}$$

$$H_2O(l) + 241.8 \text{ kJ/mol} \rightarrow H_2(g) + \tfrac{1}{2}O_2(g)$$

_____ **a.** Is energy being taken in or is it being released as liquid H_2O decomposes?

_____ **b.** What is the appropriate sign for the enthalpy change in this decomposition reaction?

PROBLEMS   Write the answer on the line to the left. Show all your work in the space provided.

3. _____ If 200. g of water at 20°C absorbs 41 840 J of energy, what will its final temperature be?

4. _____ Aluminum has a specific heat of 0.900 J/(g·°C). How much energy in kJ is needed to raise the temperature of a 625 g block of aluminum from 30.7°C to 82.1°C?

5. The products in a reaction have an enthalpy of 458 kJ/mol, and the reactants have an enthalpy of 658 kJ/mol.

_____ **a.** What is the value of $\triangle H$ for this reaction?

SECTION 1 continued

_____ **b.** Which is the more stable part of this system, the reactants or the products?

6. The enthalpy of combustion of acetylene gas is -1301.1 kJ/mol of C_2H_2.

a. Write the balanced thermochemical equation for the complete combustion of C_2H_2.

_____

_____ **b.** If 0.25 mol of C_2H_2 reacts according to the equation in part **a**, how much energy is released?

_____ **c.** How many grams of C_2H_2 are needed to react, according to the equation in part **a**, to release 3900 kJ of energy?

7. _____ Determine the $\triangle H$ for the reaction between Al and Fe_2O_3, according to the equation $2Al + Fe_2O_3 \rightarrow Al_2O_3 + 2Fe$. The enthalpy of formation of Al_2O_3 is -1676 kJ/mol. For Fe_2O_3 it is -826 kJ/mol.

8. _____ Use the data in **Appendix Table A-14** of the text to determine the $\triangle H$ for the following equation.

$$2H_2O_2(l) \rightarrow 2H_2O(l) + O_2(g)$$

CHAPTER 16 REVIEW
Reaction Energy

SECTION 2

SHORT ANSWER   Answer the following questions in the space provided.

1. For the following examples, state whether the change in entropy favors the forward or reverse reaction:

   _____ **a.** $HCl(l) \rightleftarrows HCl(g)$

   _____ **b.** $C_6H_{12}O_6(aq) \rightleftarrows C_6H_{12}O_6(s)$

   _____ **c.** $2NH_3(g) \rightleftarrows N_2(g) + 3H_2(g)$

   _____ **d.** $3C_2H_4(g) \rightleftarrows C_6H_{12}(l)$

2. _____ **a.** Write an equation that shows the relationship between enthalpy, ΔH, entropy, ΔS, and free energy, ΔG.

   _____ **b.** For a reaction to occur spontaneously, the sign of ΔG should be ____.

3. Consider the following equation: $NH_3(g) + H_2O(l) \rightleftarrows NH_4^+(aq) + OH^-(aq) + energy$

   _____ **a.** The enthalpy factor favors the forward reaction. True or False?

   _____ **b.** The sign of $T\Delta S^o$ is negative. This means the entropy factor favors the ____.

   c. Given that ΔG^o for the above reaction in the forward direction is positive, which term is greater in magnitude and therefore predominates, $T\Delta S$ or ΔH?

   _____

   _____

   _____

4. Consider the following equation for the vaporization of water:

   $$H_2O(l) \rightleftarrows H_2O(g) \qquad \Delta H = +40.65 \text{ kJ/mol at } 100°C$$

   _____ **a.** Is the forward reaction exothermic or endothermic?

   _____ **b.** Does the enthalpy factor favor the forward or reverse reaction?

   _____ **c.** Does the entropy factor favor the forward or reverse reaction?

SECTION 2 continued

PROBLEMS   Write the answer on the line to the left. Show all your work in the space provided.

5. Halogens can combine with other halogens to form several unstable compounds.
Consider the following equation: $I_2(s) + Cl_2(g) \rightleftarrows 2ICl(g)$
ΔH_f^0 for the formation of ICl = +18.0 kJ/mol and $\Delta G^0 = -5.4$ kJ/mol.

_____ **a.** Is the forward or reverse reaction favored by the enthalpy factor?

_____ **b.** Will the forward or reverse reaction occur spontaneously at standard conditions?

_____ **c.** Is the forward or reverse reaction favored by the entropy factor?

_____ **d.** Calculate the value of $T\Delta S$ for this system.

_____ **e.** Calculate the value of ΔS for this system at 25°C.

6. Calculate the free-energy change for the reactions described by the equations below. Determine whether each reaction will be spontaneous or nonspontaneous.

_____ **a.** $C(s) + 2H_2(g) \rightarrow CH_4(g)$

$\Delta S^0 = -80.7$ J/(mol•K), $\Delta H^0 = -75.0$ kJ/mol, $T = 298$ K

_____ **b.** $3Fe_2O_3(s) \rightarrow 2Fe_3O_4(s) + \frac{1}{2}O_2(g)$

$\Delta S^0 = 134.2$ J/(mol•K), $\Delta H^0 = 235.8$ kJ/mol, $T = 298$ K

CHAPTER 16 REVIEW

Reaction Energy

MIXED REVIEW

SHORT ANSWER   Answer the following questions in the space provided.

1. Describe Hess's law.

_____

_____

_____

_____

_____

2. What determines the amount of energy absorbed by a material when it is heated?

_____

_____

_____

_____

3. Describe what is meant by *enthalpy of combustion* and how a combustion calorimeter measures this enthalpy.

_____

_____

_____

_____

_____

_____

_____

MIXED REVIEW continued

4. The following equation represents a reaction that is strongly favored in the forward direction:

$$2C_7H_5(NO_2)_3(l) + 12O_2(g) \rightarrow 14CO_2(g) + 5H_2O(g) + 3N_2O(g) + energy$$

   a. Why would ΔG be negative in the above reaction?

   _____

   _____

   _____

PROBLEMS Write the answer on the line to the left. Show all your work in the space provided.

5. Consider the following equation and data: $2NO_2(g) \rightarrow N_2O_4(g)$

$$\Delta H_f^0 \text{ of } N_2O_4 = +9.2 \text{ kJ/mol}$$

$$\Delta H_f^0 \text{ of } NO_2 = +33.2 \text{ kJ/mol}$$

$$\Delta G^0 = -4.7 \text{ kJ/mol } N_2O_4$$

   _____ Use Hess's law to calculate ΔH^0 for the above reaction.

6. _____ Calculate the energy needed to raise the temperature of 180.0 g of water from 10.0°C to 40.0°C. The specific heat of water is 4.18 J/(K · g).

7. a. _____ Calculate the change in Gibbs free energy for the following equation at 25°C.

$$2H_2O_2(l) \rightarrow 2H_2O(l) + O_2(g)$$

$$\text{Given } \Delta H = -196.0 \text{ kJ/mol}$$

$$\Delta S = +125.9 \text{ J/mol}$$

   b. _____ Is this reaction spontaneous?

Name _____ Date _____ Class _____

SECTION 1

SHORT ANSWER Answer the following questions in the space provided.

1. Refer to the energy diagram below to answer the following questions.

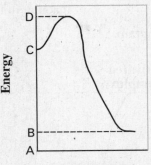

Course of reaction →

_____ **a.** Which letter represents the energy of the activated complex?

     (a) A   (c) C
     (b) B   (d) D

_____ **b.** Which letter represents the energy of the reactants?

     (a) A   (c) C
     (b) B   (d) D

_____ **c.** Which of the following represents the quantity of activation energy for the forward reaction?

     (a) the amount of energy at C minus the amount of energy at B
     (b) the amount of energy at D minus the amount of energy at A
     (c) the amount of energy at D minus the amount of energy at B
     (d) the amount of energy at D minus the amount of energy at C

_____ **d.** Which of the following represents the quantity of activation energy for the reverse reaction?

     (a) the amount of energy at C minus the amount of energy at B
     (b) the amount of energy at D minus the amount of energy at A
     (c) the amount of energy at D minus the amount of energy at B
     (d) the amount of energy at D minus the amount of energy at C

_____ **e.** Which of the following represents the energy change for the forward reaction?

     (a) the amount of energy at C minus the amount of energy at B
     (b) the amount of energy at B minus the amount of energy at C
     (c) the amount of energy at D minus the amount of energy at B
     (d) the amount of energy at B minus the amount of energy at A

Name _____ Date _____ Class _____

SECTION 1 continued

2. For the reaction described by the equation A + B → X, the activation energy for the forward direction equals 85 kJ/mol and the activation energy for the reverse direction equals 80 kJ/mol.

_____ **a.** Which has the greater energy content, the reactants or the product?

_____ **b.** What is the enthalpy of reaction in the forward direction?

_____ **c.** The enthalpy of reaction in the reverse direction is equal in magnitude but opposite in sign to the enthalpy of reaction in the forward direction. True or False?

3. Below is an incomplete energy diagram.

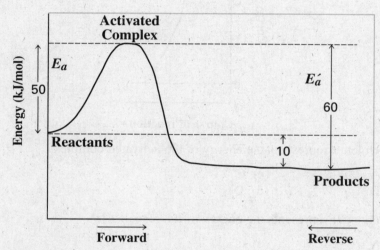

a. Use the following data to complete the diagram: E_a = +50 kJ/mol; $\Delta E_{forward}$ = −10 kJ/mol. Label the reactants, products, ΔE, E_a, E_a', and the activated complex.

_____ **b.** What is the value of E_a'?

4. It is proposed that ozone undergoes the following two-step mechanism in our upper atmosphere.

$$O_3(g) \rightarrow O_2(g) + O(g)$$
$$O_3(g) + O(g) \rightarrow 2O_2(g)$$

a. Identify any intermediates formed in the above equations.

_____

b. Write the net equation.

_____

_____ **c.** If ΔE is negative for the reaction in part **b**, what type of reaction is represented?

138   REACTION KINETICS   MODERN CHEMISTRY

Name _____ Date _____ Class _____

SECTION 2

SHORT ANSWER Answer the following questions in the space provided.

1. Below is an energy diagram for a particular process. One curve represents the energy profile for the uncatalyzed reaction, and the other curve represents the energy profile for the catalyzed reaction.

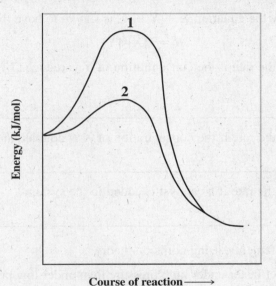

_____ **a.** Which curve has the greater activation energy?

(a) curve 1
(b) curve 2
(c) Both are equal.

_____ **b.** Which curve has the greater energy change, ΔE?

(a) curve 1
(b) curve 2
(c) Both are equal.

_____ **c.** Which curve represents the catalyzed process?

(a) curve 1
(b) curve 2

d. Explain your answer to part **c.**

_____

_____

_____

2. Is it correct to say that a catalyst affects the speed of a reaction but does not take part in the reaction? Explain your answer.

_____

_____

_____

_____

3. The reaction described by the equation $X + Y \rightarrow Z$ is shown to have the following rate law:

$$R = k[X]^3[Y]$$

a. What is the effect on the rate if the concentration of Y is reduced by one-third and [X] remains constant?

_____

b. What is the effect on the rate if the concentration of X is doubled and [Y] remains constant?

_____

c. What is the effect on the rate if a catalyst is added to the system?

_____

4. Explain the following statements, using collision theory:

a. Gaseous reactants react faster under high pressure than under low pressure.

_____

_____

b. Ionic compounds react faster when in solution than as solids.

_____

_____

c. A class of heterogeneous catalysts called surface catalysts work best as a fine powder.

_____

_____

_____

MODERN CHEMISTRY

CHAPTER 17 REVIEW
Reaction Kinetics

SHORT ANSWER   Answer the following questions in the space provided.

1. The reaction for the decomposition of hydrogen peroxide is $2H_2O_2(aq) \rightarrow 2H_2O(l) + O_2(g)$.

List three ways to speed up the rate of decomposition. For each one, briefly explain why it is effective, based on collision theory.

_____

_____

_____

_____

_____

_____

_____

2. An ingredient in smog is the gas NO. One reaction that controls the concentration of NO is

$$H_2(g) + 2NO(g) \rightarrow H_2O(g) + N_2O(g).$$

At high temperatures, doubling the concentration of H_2 doubles the rate of reaction, while doubling the concentration of NO increases the rate fourfold.

Write a rate law for this reaction consistent with these data.

_____

3. Use the following chemical equation to answer the question below:

$$Mg(s) + 2H_3O^+(aq) + Cl^-(aq) \rightarrow Mg^{2+}(aq) + 2Cl^-(aq) + H_2(g) + H_2O(l)$$

If 0.048 g of magnesium completely reacts in 20 s, what is the average reaction rate in moles/second over that time interval?

_____

_____

PROBLEMS   Write the answer on the line to the left. Show all your work in the space provided.

4. Answer the following questions using the energy diagram below.

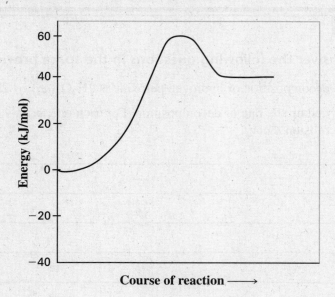

_____ **a.** Is the forward reaction represented by the curve exothermic or endothermic?

_____ **b.** Estimate the magnitude and sign of $\Delta E_{forward}$.

_____ **c.** Estimate E_a'.

A catalyst is added to the reaction, which lowers E_a by about 15 kJ/mol.

_____ **d.** Does the forward reaction rate speed up or slow down?

_____ **e.** Does the reverse reaction rate speed up or slow down?

_____ **f.** Does $\Delta E_{forward}$ change from its value in part **b**?

5. a. Determine the overall balanced equation for a reaction having the following proposed mechanism:

Step 1: $2NO + H_2$ $N_2 + H_2O_2$ Slow
Step 2: $H_2 + H_2O_2$ $2H_2O$ Fast

_____

b. Which is the rate-determining step?

_____

c. What is the intermediate in the above reaction?

_____

CHAPTER 18 REVIEW
Chemical Equilibrium

SECTION 1

SHORT ANSWER   Answer the following questions in the space provided.

1. Write the equilibrium expression for the following hypothetical equation:

$$3A(aq) + B(aq) \rightleftarrows 2C(aq) + 3D(aq)$$

2. a. Write the appropriate chemical equilibrium expression for each of the following equations. Include the value of K.

   (1) $N_2O_4(g) \rightleftarrows 2NO_2(g)$ $K = 0.1$

   (2) $NH_4OH(aq) \rightleftarrows NH_4^+(aq) + OH^-(aq)$ $K = 2 \times 10^{-5}$

   (3) $H_2(g) + I_2(g) \rightleftarrows 2HI(g)$ $K = 54.0$

   (4) $2SO_2(g) + O_2(g) \rightleftarrows 2SO_3(g)$ $K = 1.8 \times 10^{-2}$

Name _____ Date _____ Class _____

_____ b. Which of the four systems in part **a** proceeds furthest forward before equilibrium is established?

_____ c. Which system contains mostly reactants at equilibrium?

3. a. Compare the rates of forward and reverse reactions when equilibrium has been reached.

_____

_____

b. Describe what happens to the concentrations of reactants and products when chemical equilibrium has been reached.

_____

_____

PROBLEMS Write the answer on the line to the left. Show all your work in the space provided.

4. _____ Consider the following equation:

$$2NO(g) + O_2(g) \rightleftarrows 2NO_2(g)$$

At equilibrium, $[NO] = 0.80$ M, $[O_2] = 0.50$ M, and $[NO_2] = 0.60$ M. Calculate the value of K for this reaction.

5. _____ What is the K value for the following equation if the gaseous mixture in a 4 L container reaches equilibrium at 1000 K and contains 4.0 mol of N_2, 6.4 mol of H_2, and 0.40 mol of NH_3?

$$N_2(g) + 3H_2(g) \rightleftarrows 2NH_3(g)$$

CHAPTER 18 REVIEW
Chemical Equilibrium

SECTION 2

SHORT ANSWER Answer the following questions in the space provided.

1. _____ Raising the temperature of any equilibrium system always

    (a) favors the forward reaction.
    (b) favors the reverse reaction.
    (c) favors the exothermic reaction.
    (d) favors the endothermic reaction.

2. Consider the following equilibrium equation: $CH_3OH(g) + 101$ kJ $\rightleftarrows CO(g) + 2H_2(g)$.

    _____ **a.** Increasing [CO] will

    (a) increase $[H_2]$.   (c) not change $[H_2]$.
    (b) decrease $[H_2]$.   (d) cause $[H_2]$ to fluctuate.

    _____ **b.** Raising the temperature will cause the equilibrium of the system to

    (a) favor the reverse reaction.   (c) shift back and forth.
    (b) favor the forward reaction.   (d) remain as it was before.

    _____ **c.** Raising the temperature will

    (a) increase the value of K.
    (b) decrease the value of K.
    (c) not change the value of K.
    (d) make the value of K fluctuate.

3. Consider the following equilibrium equation: $H_2O(g) + C(s) \rightleftarrows H_2(g) + CO(g) +$ energy
At equilibrium, which reaction will be favored (forward, reverse, or neither) when

    _____ **a.** extra CO gas is introduced?

    _____ **b.** a catalyst is introduced?

    _____ **c.** the temperature of the system is lowered?

    _____ **d.** the pressure on the system is increased due to a decrease in the container volume?

4. _____ Silver chromate dissolves in water according to the following equation:

$$Ag_2CrO_4(s) \rightleftarrows 2Ag^+(aq) + CrO_4^{2-}(aq)$$

Which of these correctly represents the equilibrium expression for the above equation?

    (a) $\dfrac{2[Ag^+] + [CrO_4^{2-}]}{Ag_2CrO_4}$ (b) $\dfrac{[Ag_2CrO_4]}{[Ag^+]^2[CrO_4^{2-}]}$ (c) $\dfrac{[Ag^+]^2[CrO_4^{2-}]}{1}$ (d) $\dfrac{[Ag^+]^2[CrO_4^{2-}]}{2[Ag_2CrO_4]}$

SECTION 2 continued

5. Are pure solids included in equilibrium expressions? Explain your answer.

_____

_____

_____

6. A key step in manufacturing sulfuric acid is represented by the following equation:

$$2SO_2(g) + O_2(g) \rightleftarrows 2SO_3(g) + 100 \text{ kJ/mol}$$

To be economically viable, this process must yield as much SO_3 as possible in the shortest possible time. You are in charge of this manufacturing process.

a. Would you impose a high pressure or a low pressure on the system? Explain your answer.

_____

_____

b. To maximize the yield of SO_3, should you keep the temperature high or low during the reaction?

_____

_____

c. Will adding a catalyst change the yield of SO_3?

_____

_____

7. The equation for an equilibrium system easily studied in a lab follows:

$$2NO_2(g) \rightleftarrows N_2O_4(g)$$

N_2O_4 gas is colorless, and NO_2 gas is dark brown. Lowering the temperature of the equilibrium mixture of gases reduces the intensity of the color.

a. Is the forward or reverse reaction favored when the temperature is lowered?

_____

_____

b. Will the sign of ΔH be positive or negative if the temperature is lowered? Explain your answer.

_____

_____

_____

CHAPTER 18 REVIEW
Chemical Equilibrium

SECTION 3

SHORT ANSWER Answer the following questions in the space provided.

1. _____ Lime juice turns litmus paper red, indicating that lime juice is
   (a) acidic.
   (b) basic.
   (c) neutral.
   (d) alkaline

2. _____ Addition of the salt of a weak acid to a solution of the weak acid
   (a) lowers the concentration of the nonionized acid and the concentration of the H_3O^+ ion.
   (b) lowers the concentration of the nonionized acid and raises the concentration of the H_3O^+ ion.
   (c) raises the concentration of the nonionized acid and the concentration of the H_3O^+ ion.
   (d) raises the concentration of the nonionized acid and lowers the concentration of the H_3O^+ ion.

3. _____ Salts of a weak acid and a strong base produce solutions that are
   (a) acidic only.
   (b) basic only.
   (c) neutral only.
   (d) either acidic, basic, or neutral.

4. _____ If an acid is added to a solution of a weak base and its salt,
   (a) more water is formed and more weak base ionizes.
   (b) hydronium ion concentration decreases.
   (c) more hydroxide ion is formed.
   (d) more nonionized weak base is formed.

5. **a.** In the space below each of the following equations, correctly label the two conjugate acid-base pairs as *acid 1, acid 2, base 1,* and *base 2.*
   (a) $CO_3^{2-}(aq) + H_3O^+(aq) \rightleftarrows HCO_3^-(aq) + H_2O(l)$

   (b) $HPO_4^{2-}(aq) + H_2O(l) \rightleftarrows OH^-(aq) + H_2PO_4^-(aq)$

   _____ **b.** Which reaction in part **a** is an example of hydrolysis?

   _____ **c.** As the first reaction in part **a** proceeds, the pH of the solution
   (a) increases.   (c) stays at the same level.
   (b) decreases.   (d) fluctuates.

SECTION 3 continued

6. Write the formulas for the acid and the base that could form the salt $Ca(NO_3)_2$.

_____

7. Consider the following equation for the reaction of a weak base in water:

$$NH_3(aq) + H_2O(l) \rightleftarrows NH_4^+(aq) + OH^-(aq)$$

Write the equilibrium expression for K.

PROBLEMS Write the answer on the line to the left. Show all your work in the space provided.

8. An unknown acid X hydrolyzes according to the equation in part **a** below.

a. In the space below the equation, correctly label the two conjugate acid-base pairs in this system as *acid 1, acid 2, base 1,* and *base 2.*

$$HX(aq) + H_2O(l) \rightleftarrows X^-(aq) + H_3O^+(aq)$$

b. Write the equilibrium expression for K_a for this system.

_____ c. Experiments show that at equilibrium $[H_3O^+] = [X^-] = 2.0 \times 10^{-5}$ mol/L and $[HX] = 4.0 \times 10^{-2}$ mol/L. Calculate the value of K_a based on these data.

CHAPTER 18 REVIEW
Chemical Equilibrium

SECTION 4

SHORT ANSWER **Answer the following questions in the space provided.**

1. Match the solution type on the right to the corresponding relationship between the ion product and the K_{sp} for that solution, listed on the left.

   _____ The ion product exceeds the K_{sp}.

   _____ The ion product equals the K_{sp}.

   _____ The ion product is less than the K_{sp}.

   (a) The solution is saturated; no more solid will dissolve.

   (b) The solution is unsaturated; no solid is present.

   (c) The solution is supersaturated; solid may form if the solution is disturbed.

2. Silver carbonate, Ag_2CO_3, makes a saturated solution with $K_{sp} = 10^{-11}$.

   a. Write the equilibrium expression for the dissolution of Ag_2CO_3.

   _____ **b.** In this system, will the foward or reverse reaction be favored if extra Ag^+ ions are added?

PROBLEMS **Write the answer on the line to the left. Show all your work in the space provided.**

3. When the ionic solid XCl_2 dissolves in pure water to make a saturated solution, experiments show that 2×10^{-3} mol/L of X^{2+} ions go into solution.

   _____ **a.** Write the equation showing the dissolution of XCl_2 and the corresponding equilibrium expression.

   _____

   _____ **b.** Calculate the value of K_{sp} for XCl_2.

Name _____ Date _____ Class _____

_____ **c.** Refer to **Table 3** on page 615 of the text. Would XCl_2 be more soluble or less soluble than $PbCl_2$ at the same temperature?

4. The solubility of Ag_3PO_4 is 2.1×10^{-4} g/100. g.

_____ **a.** Write the equation showing the dissolution of this ionic solid.

_____ **b.** Calculate the molarity of this saturated solution.

_____ **c.** What is the value of K_{sp} for this system?

5. As $PbCl_2$ dissolves, $[Pb^{2+}] = 2.0 \times 10^{-1}$ mol/L and $[Cl^-] = 1.5 \times 10^{-2}$ mol/L.

_____ **a.** Write the equilibrium expression for the dissolution of $PbCl_2$.

_____ **b.** Compute the ion product, using the data given above.

CHAPTER 18 REVIEW
Chemical Equilibrium

MIXED REVIEW

SHORT ANSWER   Answer the following questions in the space provided.

1. Consider the following equilibrium equation:

$$N_2(g) + 2O_2(g) \rightleftarrows 2NO_2(g); \Delta H = +33 \text{ kJ/mol}$$

At equilibrium, which reaction is favored when

_____   **a.** some N_2 is removed?

_____   **b.** a catalyst is introduced?

_____   **c.** pressure on the system increases due to a decrease in the volume?

_____   **d.** the temperature of the system is increased?

2. Ammonia gas dissolves in water according to the following equation:

$$NH_3(g) + H_2O(l) \rightleftarrows NH_4^+(aq) + OH^-(aq) + \text{energy}; K = 1.8 \times 10^{-5}$$

_____   **a.** Is aqueous ammonia an acid or a base?

_____   **b.** Is the equation given above an example of hydrolysis?

_____   **c.** For the given value of K, does the equilibrium favor the forward or reverse reaction?

PROBLEMS   Write the answer on the line to the left. Show all your work in the space provided.

3. Formic acid, HCOOH, is a weak acid present in the venom of red-ants. At equilibrium, [HCOOH] = 2.00 M, [HCOO$^-$] = 4.0×10^{-1} M, and [H$_3$O$^+$] = 9.0×10^{-4} M.

_____   **a.** Write the equilibrium expression for the ionization of formic acid.

_____   **b.** Calculate the value of K_a for this acid.

4. HF hydrolyzes according to the following equation:

$$HF(aq) + H_2O(l) \rightleftarrows H_3O^+(aq) + F^-(aq)$$

When 0.0300 mol of HF dissolves in 1.00 L of water, the solution quickly ionizes to reach equilibrium. At equilibrium, the remaining [HF] = 0.0270 M.

_____ **a.** How many moles of HF ionize per liter of water to reach equilibrium?

_____ **b.** What are [F$^-$] and [H$_3$O$^+$]?

_____ **c.** What is the value of K_a for HF?

5. Refer to **Table 3** on page 615 of the text. CaSO$_4$(s) is only slightly soluble in water.

_____ **a.** Write the equilibrium equation and equilibrium expression for the dissolution of CaSO$_4$(s) with the K_{sp} value.

_____ **b.** Determine the solubility of CaSO$_4$ at 25°C in grams per 100. g H$_2$O.

CHAPTER 19 REVIEW
Oxidation-Reduction Reactions

SECTION 1

SHORT ANSWER   Answer the following questions in the space provided.

1. _____   All the following equations involve redox reactions *except*

    (a) $CaO + H_2O \rightarrow Ca(OH)_2$.
    (b) $2SO_2 + O_2 \rightarrow 2SO_3$.
    (c) $2HgO \rightarrow 2Hg + O_2$.
    (d) $SnCl_4 + 2FeCl_2 \rightarrow 2FeCl_3 + SnCl_2$.

2. Assign the correct oxidation number to the individual atom or ion below.

_____   **a.** Mn in MnO_2

_____   **b.** S in S_8

_____   **c.** Cl in $CaCl_2$

_____   **d.** I in IO_3^-

_____   **e.** C in HCO_3^-

_____   **f.** Fe in $Fe_2(SO_4)_3$

_____   **g.** S in $Fe_2(SO_4)_3$

3. In each of the following half-reactions, determine the value of x.

_____   **a.** $S^{6+} + x\,e^- \rightarrow S^{2-}$

_____   **b.** $2Br^x \rightarrow Br_2 + 2e^-$

_____   **c.** $Sn^{4+} + 2e^- \rightarrow Sn^x$

_____   **d.** Which of the above half-reactions represent reduction processes?

4. Give examples, other than those listed in **Table 1** on page 631 of the text, for the following:

_____   **a.** a compound containing H in a -1 oxidation state

_____   **b.** a peroxide

_____   **c.** a polyatomic ion in which the oxidation number for S is $+4$

_____   **d.** a substance in which the oxidation number for F is not -1

5. OILRIG is a mnemonic device often used by students to help them understand redox reactions.

"Oxidation is loss, reduction is gain."

Explain what that phrase means—loss and gain of what?

_____

_____

6. For each of the reactions described by the following equations, state whether or not any oxidation and reduction is occurring, and write the oxidation-reduction half-reactions for those cases in which redox does occur.

a. $Ca(OH)_2(aq) + 2HCl(aq) \rightarrow CaCl_2(aq) + 2H_2O(l)$

_____

b. $CH_4(g) + 2O_2(g) \rightarrow CO_2(g) + 2H_2O(g)$

_____

c. $2Al(s) + 3CuCl_2(aq) \rightarrow 2AlCl_3(aq) + 3Cu(s)$

_____

7. I^- is converted into I_2 by the addition of an aqueous solution of KMnO4 to an aqueous solution of KI.

a. What is the oxidation number assigned to I_2?

_____

b. The conversion of I^- to I_2 is a(n) _____ reaction.

_____

c. How many electrons are lost when 1 mol of I_2 is formed from I^-?

_____

_____

CHAPTER 19 REVIEW

Oxidation-Reduction Reactions

SECTION 2

SHORT ANSWER   Answer the following questions in the space provided.

1. _____ All of the following should be done in the process of balancing redox equations *except*

    (a) adjusting coefficients to balance atoms.

    (b) adjusting coefficients in electron equations to balance numbers of electrons lost and gained.

    (c) adjusting subscripts to balance atoms.

    (d) writing two separate electron equations.

2. MnO_4^- can be reduced to MnO_2.

_____ **a.** Assign the oxidation number to Mn in these two species.

_____ **b.** How many electrons are gained per Mn atom in this reduction?

_____ **c.** If 0.50 mol of MnO_4^- is reduced, how many electrons are gained?

3. Iodide ions can be oxidized to form iodine. Write the balanced oxidation half-reaction for the oxidation of iodide to iodine.

_____

4. Some bleaches contain aqueous chlorine as the active ingredient. Aqueous chlorine is made by dissolving chlorine gas in water. Aqueous chlorine is capable of oxidizing iron(II) ions to iron(III) ions. When iron(II) ions are oxidized, chloride ions are formed.

    a. Write equations for the two half-reactions involved. Label them *oxidation* or *reduction*.

_____

_____

    b. Write the balanced ionic equation for the redox reaction between aqueous chlorine and iron(II).

_____

    c. Show that the equation in part **b** is balanced by charge.

_____

_____

SECTION 2 continued

5. Write the equations for the oxidation and reduction half-reactions for the redox reactions below, and then balance the reaction equations.

  a. $MnO_2(s) + HCl(aq) \rightarrow MnCl_2(aq) + Cl_2(g) + H_2O(l)$

  b. $S(s) + HNO_3(aq) \rightarrow SO_3(g) + H_2O(l) + NO_2(g)$

  c. $H_2C_2O_4(aq) + K_2CrO_4(aq) + HCl(aq) \rightarrow CrCl_3(aq) + KCl(aq) + H_2O(l) + CO_2(g)$

CHAPTER 19 REVIEW
Oxidation-Reduction Reactions

SECTION 3

SHORT ANSWER   Answer the following questions in the space provided.

1. For each of the following, identify the stronger oxidizing or reducing agent. (Refer to **Table 3** on page 643 of the text.)

_____ **a.** Ca or Cu as a reducing agent

_____ **b.** Ag^+ or Na^+ as an oxidizing agent

_____ **c.** Fe^{3+} or Fe^{2+} as an oxidizing agent

2. For each of the following incomplete equations, state whether a redox reaction is likely to occur. (Refer to **Table 3** on page 643 of the text.)

_____ **a.** $Mg + Sn^{2+} \rightarrow$

_____ **b.** $Ag + Cu^{2+} \rightarrow$

_____ **c.** $Br_2 + I^- \rightarrow$

3. Label each of the following statements about redox as True or False.

_____ **a.** A strong oxidizing agent is itself readily reduced.

_____ **b.** In disproportionation, one chemical acts as both an oxidizing agent and a reducing agent in the same process.

_____ **c.** The number of moles of chemical oxidized must equal the number of moles of chemical reduced.

4. Solutions of Fe^{2+} are fairly unstable, in part because they can undergo disproportionation, as shown by the following unbalanced equation:

$$Fe^{2+} \rightarrow Fe^{3+} + Fe$$

a. Balance the above equation.

_____

_____ **b.** If the reaction described above produces 0.036 mol of Fe, how many moles of Fe^{3+} will form?

SECTION 3 continued

5. Oxygen gas is a powerful oxidizing agent.

_____ **a.** Assign the oxidation number to O_2.

_____ **b.** What does oxygen's oxidation number usually become when it functions as an oxidizing agent?

c. Approximately where would you place O_2 in the list of oxidizing agents in **Table 3** on page 643 of the text?

_____

d. Describe the changes in oxidation states that occur in carbon and oxygen, and identify the oxidizing and reducing agents, in the combustion reaction described by the following equation:

$$C_6H_{12}O_6(s) + 6O_2(g) \rightarrow 6CO_2(g) + 6H_2O(l)$$

_____

_____

_____

_____

6. An example of disproportionation is the slow decomposition of aqueous chlorine, $Cl_2(aq)$, represented by the following unbalanced equation:

$$Cl_2(aq) + H_2O(l) \rightarrow ClO^-(aq) + Cl^-(aq) + H^+(aq)$$

a. Show that the oxygen and hydrogen atoms in the above reaction are not changing oxidation states.

_____

_____

b. Show the changes in the oxidation states of chlorine as this reaction proceeds.

_____

_____ **c.** In the oxidation reaction, how many electrons are transferred per Cl atom?

_____ **d.** In the reduction reaction, how many electrons are transferred per Cl atom?

e. What must be the ratio of ClO^- to Cl^- in the above reaction? Explain your answer.

_____

_____

f. Balance the equation for the decomposition of $Cl_2(aq)$.

_____

Name _____ Date _____ Class _____

CHAPTER 19 REVIEW
Oxidation-Reduction Reactions

MIXED REVIEW

SHORT ANSWER   Answer the following questions in the space provided.

1. Label the following descriptions of reactions *oxidation, reduction,* or *disproportionation.*

_____   **a.** conversion of Na_2O_2 to NaO and O_2

_____   **b.** conversion of Br^- to Br_2

_____   **c.** conversion of Fe^{2+} to Fe^{3+}

_____   **d.** a change in the oxidation number in a negative direction

PROBLEMS   Write the answer on the line to the left. Show all your work in the space provided.

2. Consider the following unbalanced equation:

$$KMnO_4(aq) + HCl(aq) + Al(s) \rightarrow AlCl_3(aq) + MnCl_2(aq) + KCl(aq) + H_2O(l)$$

a. Write the oxidation and reduction half-reactions. Label each half-reaction *oxidation* or *reduction.*

_____

_____

b. Balance the equation using the seven-step procedure shown on pages 637–639 of the text.

_____   **c.** Identify the oxidizing agent in this system.

MIXED REVIEW continued

3. Consider the unbalanced ionic equation $ClO^- + H^+ \rightarrow Cl_2 + ClO_3^- + H_2O$.

    a. Assign the oxidation number to each element.

_____

_____

_____

_____

_____

_____   **b.** How many electrons are given up by each Cl atom as it is oxidized?

_____   **c.** How many electrons are gained by each Cl atom as it is reduced?

_____   **d.** Is this an example of disproportionation?

    e. Balance the above equation, using the method of your choice.

CHAPTER 20 REVIEW

Electrochemistry

SECTION 1

SHORT ANSWER   Answer the following questions in the space provided.

1. The following reaction takes place in an electrochemical cell:

$$Cu^{2+}(aq) + Zn(s) \rightarrow Cu(s) + Zn^{2+}(aq)$$

a. Which electrode is the anode?

_____

_____

b. Which electrode is the cathode?

_____

_____

c. Write the cell notation for this system.

_____

_____

d. Write equations for the half-reactions that occur at each electrode and label each reaction.

_____

_____

2. _____ Energy will be released in the form of heat when

    (a) reactants in a spontaneous energy-releasing redox reaction are connected by a wire.
    (b) reactants in a spontaneous energy-releasing redox reaction are in direct contact.
    (c) copper atoms are deposited on an anode.
    (d) electrochemical half-cells are isolated from one another.

3. _____ An electrochemical cell is constructed using the reaction of chromium metal and iron(II) ion, as follows:

$$2Cr(s) + 3Fe^{2+}(aq) \rightarrow 2Cr^{3+} + 3Fe(s)$$

Which statement best describes this system?

    (a) Electrons flow from the iron electrode to the chromium electrode.
    (b) Energy is released.
    (c) Negative ions move through the salt bridge from the chromium half-cell to the iron half-cell.
    (d) Negative ions move through the salt bridge from the iron half-cell to the chromium half-cell.

SECTION 1 continued

4. Below is a diagram of an electrochemical cell.

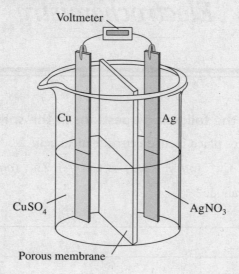

Voltmeter

Cu   Ag

CuSO$_4$ AgNO$_3$

Porous membrane

a. Write the anode half-reaction.

_____

b. Write the cathode half-reaction.

_____

c. Write the balanced overall cell reaction.

_____

_____ **d.** Do electrons within the electrochemical cell travel through the voltmeter in a clockwise or a counterclockwise direction, as represented in the diagram?

_____ **e.** In what direction do anions pass through the porous membrane, as represented in the diagram?

CHAPTER 20 REVIEW

Electrochemistry

SECTION 2

SHORT ANSWER Answer the following questions in the space provided.

1. _____ In a voltaic cell, transfer of charge through the external wires occurs by means of

    (a) ionization.

    (b) ion movement.

    (c) electron movement.

    (d) proton movement.

2. _____ All the following claims about voltaic cells are true *except*

    (a) E^o_{cell} is positive.

    (b) The redox reaction in the cell occurs without the addition of electric energy.

    (c) Electrical energy is converted to chemical energy.

    (d) Chemical energy is converted to electrical energy.

3. Use **Table 1** on page 664 of the text to find E^o for the following:

_____ **a.** the reduction of MnO_4^{1-} to MnO_4^{2-}

_____ **b.** the oxidation of Cr to Cr^{3+}

_____ **c.** the reaction within the SHE

_____ **d.** $Cl_2 + 2Br^- \rightarrow 2Cl^- + Br_2$

4. Why does a zinc coating protect steel from corrosion?

_____

_____

5. Which two types of batteries share the following anode half-reaction?

$$Zn(s) + 2OH^-(aq) \rightarrow Zn(OH)_2(s) + 2e^-$$

_____

_____

6. Complete the following sentences:

Corrosion acts as a voltaic cell because oxidation and reduction reactions occur

_____ at different sites. The two half-cells are connected by a

_____, which allows electrons to flow.

SECTION 2 continued

PROBLEM Write the answer on the line to the left. Show all your work in the space provided.

7. Use **Table 1** on p. 664 of the text to find E^0 for the following voltaic cells:

_____ **a.** $Al(s) \mid Al^{3+}(aq) \parallel Cd^{2+}(aq) \mid Cd(s)$

_____ **b.** $Fe(s) \mid Fe^{2+}(aq) \parallel Pb^{2+}(aq) \mid Pb(s)$

_____ **c.** $6I^-(aq) + 2Au^{3+}(aq) \rightarrow 3I_2(s) + 2Au(s)$

8. A voltmeter connected to a copper-silver voltaic cell reads +0.46 V. The silver is then replaced with metal X and its 2+ ion. A new voltage reading shows that the direction of the current has reversed, and the voltmeter reads +0.74 V. Use data from **Table 1** on page 664 of the text to answer.

_____ **a.** Calculate the reduction potential of metal X.

_____ **b.** Predict the identity of metal X.

CHAPTER 20 REVIEW
Electrochemistry

SECTION 3

SHORT ANSWER Answer the following questions in the space provided.

1. Label each of the following statements as applying to a *voltaic cell,* an *electrolytic cell,* or *both*:

_____ **a.** The cell reaction involves oxidation and reduction.

_____ **b.** The cell reaction proceeds spontaneously.

_____ **c.** The cell reaction is endothermic.

_____ **d.** The cell reaction converts chemical energy into electrical energy.

_____ **e.** The cell reaction converts electrical energy into chemical energy.

_____ **f.** The cell contains both a cathode and an anode.

2. _____ In an electrolytic cell, oxidation takes place

    (a) at the anode.
    (b) at the cathode.
    (c) via the salt bridge.
    (d) at the positive electrode.

3. _____ Which atom forms an ion that would always migrate toward the cathode in an electrolytic cell?

    (a) F
    (b) I
    (c) Cu
    (d) Cl

4. An electrolytic process in which solid metal is deposited on a surface is called _____.

5. When a rechargeable camera battery is being recharged, the cell acts as a(n) _____ cell and converts _____ energy into _____ energy. When the battery is being used to power the camera, it acts as a(n) _____ cell and converts _____ energy into _____ energy.

SECTION 3 continued

6. Using **Figure 14** on p. 668 as a guide, describe how you would electroplate gold, Au, onto a metal object from a solution of $Au(CN)_3$. Include in your discussion, the equation for the reaction that plates the gold.

_____

_____

_____

_____

7. Explain why aluminum recycling is less expensive than the extraction of aluminum metal from bauxite ore.

_____

_____

_____

_____

8. Label the following statements about the electrolysis of water as True or False.

_____ The process is spontaneous.

_____ Hydrogen gas is formed at the cathode.

_____ Oxygen gas is formed at the anode.

_____ Electric energy is generated.

CHAPTER 20 REVIEW

Electrochemistry

MIXED REVIEW

SHORT ANSWER Answer the following questions in the space provided.

1. _____ An electrochemical cell consists of two electrodes separated by a(n)

  (a) anode.
  (b) cathode.
  (c) voltage.
  (d) electrolyte.

2. _____ When a car battery is charging,

  (a) electrical energy is converted into energy of motion.
  (b) energy of motion is converted into electrical energy.
  (c) chemical energy is converted into electrical energy.
  (d) electrical energy is converted into chemical energy.

3. _____ Electroplating is an application of

  (a) electrolytic cell reactions.
  (b) fuel cell reactions.
  (c) auto-oxidation reactions.
  (d) galvanic reactions.

4. _____ A major benefit of electroplating is that it

  (a) increases concentrations of toxic wastes.
  (b) protects metals from corrosion.
  (c) saves time.
  (d) leads to a buildup of impurities.

5. _____ The transfer of charge through the electrolyte solution occurs by means of

  (a) ionization.
  (b) ion movement.
  (c) electron movement.
  (d) proton movement.

MIXED REVIEW continued

6. A voltaic cell is constructed that reacts according to the following equation:

$$Mg(s) + 2H^+(aq) \rightarrow Mg^{2+} + H_2(g)$$

a. Write equations for the half-reactions that occur in this cell.

_____

_____

b. Which half-reaction occurs in the anode half-cell?

_____

_____

c. Write the cell notation for this cell.

_____

_____

d. Electrons flow through the wire from the _____ electrode to the _____ electrode. Positive ions move from the _____ half-cell to the _____ half-cell.

PROBLEM   Write the answer on the line to the left. Show all your work in the space provided.

_____ 7. Refer to **Table 1** on page 664 of the text. What is the voltage of the cell for the following reaction?

$$Mg + Ni(NO_3)_2 \rightarrow Ni + Mg(NO_3)_2$$

CHAPTER 21 REVIEW

Nuclear Chemistry

SECTION 1

SHORT ANSWER   Answer the following questions in the space provided.

1. _____ Based on the information about the three elementary particles on page 683 of the text, which has the greatest mass?

  (a) the proton
  (b) the neutron
  (c) the electron
  (d) They all have the same mass.

2. _____ The force that keeps nucleons together is

  (a) a strong nuclear force.
  (b) a weak nuclear force.
  (c) an electromagnetic force.
  (d) a gravitational force.

3. _____ The stability of a nucleus is most affected by the

  (a) number of neutrons.
  (b) number of protons.
  (c) number of electrons.
  (d) ratio of neutrons to protons.

4. _____ If an atom should form from its constituent particles,

  (a) matter is lost and energy is taken in.
  (b) matter is lost and energy is released.
  (c) matter is gained and energy is taken in.
  (d) matter is gained and energy is released.

5. _____ For atoms of a given mass number, those with greater mass defects, have

  (a) smaller binding energies per nucleon.
  (b) greater binding energies per nucleon.
  (c) the same binding energies per nucleon as those with smaller mass defects.
  (d) variable binding energies per nucleon.

6. Based on **Figure 1** on page 684 of the text, which isotope of He, helium-3 or helium-4,

  _____ **a.** has the smaller binding energy per nucleon?

  _____ **b.** is more stable to nuclear changes?

7. The number of neutrons in an atom of magnesium-25 is _____.

8. Nuclides of the same element have the same _____.

SECTION 1 continued

9. Atom X has 50 nucleons and a binding energy of 4.2×10^{-11} J. Atom Z has 80 nucleons and a binding energy of 8.4×10^{-11} J.

_____ **a.** The mass defect of Z is twice that of X. True or False?

_____ **b.** Which atom has the greater binding energy per nucleon?

_____ **c.** Which atom is likely to be more stable to nuclear transmutations?

10. Identify the missing term in each of the following nuclear equations. Write the element's symbol, its atomic number, and its mass number.

_____ **a.** $^{14}_{6}C \rightarrow {}^{0}_{-1}e +$ _____

_____ **b.** $^{63}_{29}Cu + {}^{1}_{1}H \rightarrow$ _____ $+ {}^{4}_{2}He$

11. Write the equation that shows the equivalency of mass and energy.

_____

12. Consider the two nuclides $^{56}_{26}Fe$ and $^{14}_{6}C$.

a. Determine the number of protons in each nucleus.

_____

b. Determine the number of neutrons in each nucleus.

_____

c. Determine whether the $^{56}_{26}Fe$ nuclide is likely to be stable or unstable, based on its position in the band of stability shown in **Figure 2** on page 685 of the text.

_____

PROBLEM Write the answer on the line to the left. Show all your work in the space provided.

13. _____ Neon-20 is a stable isotope of neon. Its actual mass has been found to be 19.992 44 amu. Determine the mass defect in this nuclide.

CHAPTER 21 REVIEW
Nuclear Chemistry

SECTION 2

SHORT ANSWER   Answer the following questions in the space provided.

1. _____ The nuclear equation $^{210}_{84}Po \rightarrow ^{206}_{82}Pb + ^{4}_{2}He$ is an example of an equation that represents
   (a) alpha emission.
   (b) beta emission.
   (c) positron emission.
   (d) electron capture.

2. _____ When $^{a}_{b}Z$ undergoes electron capture to form a new element X, which of the following best represents the product formed?

   (a) $^{a-1}_{b}X$
   (b) $^{a+1}_{b}X$
   (c) $^{a}_{b+1}X$
   (d) $^{a}_{b-1}X$

3. _____ Which of the following best represents the fraction of a radioactive sample that remains after four half-lives have occurred?

   (a) $\left(\dfrac{1}{2}\right)^4$ **(c)** $\left(\dfrac{1}{4}\right)$

   (b) $\left(\dfrac{1}{2}\right) \times 4$ **(d)** $\left(\dfrac{1}{4}\right)^2 \times 4$

4. Match the nuclear symbol on the right to the name of the corresponding particle on the left.

   _____ beta particle   **(a)** $^{1}_{1}p$

   _____ proton   **(b)** $^{4}_{2}He$

   _____ positron   **(c)** $^{0}_{-1}\beta$

   _____ alpha particle   **(d)** $^{0}_{+1}\beta$

5. Label each of the following statements as True or False. Consider the U-238 decay series on page 692 of the text. For the series of decays from U-234 to Po-218, each nuclide

   _____ **a.** shares the same atomic number

   _____ **b.** differs in mass number from others by multiples of 4

   _____ **c.** has a unique atomic number

   _____ **d.** differs in atomic number from others by multiples of 4

Name _____ Date _____ Class _____

SECTION 2 continued

6. _____ Identify the missing term in the following nuclear equation. Write the element's symbol, its atomic number, and its mass number.

$$\underline{\quad?\quad} \rightarrow {}^{231}_{90}Th + {}^{4}_{2}He$$

7. Lead-210 undergoes beta emission. Write the nuclear equation showing this transmutation.

_____

8. Einsteinium is a transuranium element. Einsteinium-247 can be prepared by bombarding uranium-238 with nitrogen-14 nuclei, releasing several neutrons, as shown by the following equation:

$${}^{238}_{92}U + {}^{14}_{7}N \rightarrow {}^{247}_{99}Es + x\,{}^{1}_{0}n$$

What must be the value of x in the above equation? Explain your reasoning.

_____

_____

_____

PROBLEMS Write the answer on the line to the left. Show all your work in the space provided.

9. _____ Phosphorus-32 has a half-life of 14.3 days. How many days will it take for a sample of phosphorus-32 to decay to one-eighth its original amount?

10. _____ Iodine-131 has a half-life of 8.0 days. How many grams of an original 160 mg sample will remain after 40 days?

11. _____ Carbon-14 has a half-life of 5715 years. It is used to determine the age of ancient objects. If a sample today contains 0.060 mg of carbon-14, how much carbon-14 must have been present in the sample 11 430 years ago?

CHAPTER 21 REVIEW
Nuclear Chemistry

SECTION 3

SHORT ANSWER   Answer the following questions in the space provided.

1. _____ The radioisotope cobalt-60 is used for all of the following applications *except*

    (a) killing food-spoiling bacteria.　　**(c)** treating heart disease.
    (b) preserving food.　　　　　　　　**(d)** treating certain kinds of cancers.

2. _____ All of the following contribute to background radiation exposure *except*

    (a) radon in homes and buildings.
    (b) cosmic rays passing through the atmosphere.
    (c) consumption of irradiated foods.
    (d) minerals in Earth's crust.

3. _____ Which one of the graphs shown below best illustrates the decay of a sample of carbon-14? Assume each division on the time axis represents 5715 years.

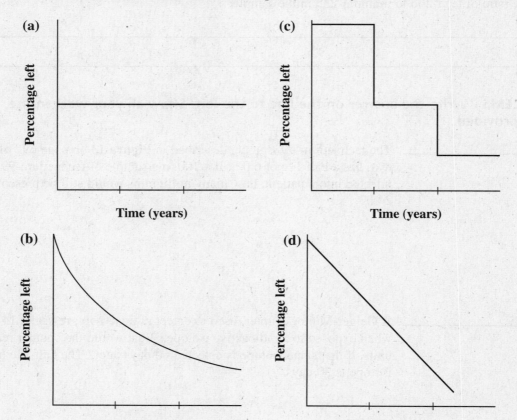

4. Match the item on the left with its description on the right.

_____ Geiger-Müller counter   **(a)** device that uses film to measure the approximate radiation exposure of people working with radiation

_____ scintillation counter   **(b)** instrument that converts scintillating light to an electric signal for detecting radiation

_____ film badge   **(c)** meter that detects radiation by counting electric pulses carried by gas ionized by radiation

_____ radioactive tracers   **(d)** radioactive atoms that are incorporated into substances so that movement of the substances can be followed by detectors

5. Which type of radiation is easiest to shield? Why?

_____

_____

6. One technique for dating ancient rocks involves uranium-235, which has a half-life of 710 million years. Rocks originally rich in uranium-235 will contain small amounts of its decay series, including the nonradioactive lead-206. Explain the relationship between a sample's relative age and the ratio of lead-206 to uranium-235 in the sample.

_____

_____

PROBLEMS Write the answer on the line to the left. Show all your work in the space provided.

7. _____ The technetium-99 isotope, described in **Figure 13** on page 697 of the text, has a half-life of 6.0 h. If a 100. mg sample of technetium-99 were injected into a patient, how many milligrams would still be present after 24 h?

8. _____ A Geiger-Müller counter, used to detect radioactivity, registers 14 units when exposed to a radioactive isotope. What would the counter read, in units, if that same isotope is detected 60 days later? The half-life of the isotope is 30 days.

CHAPTER 21 REVIEW
Nuclear Chemistry

SECTION 4

SHORT ANSWER   Answer the following questions in the space provided.

1. Match each of the following statements with the process(es) to which they apply, using one of the choices below:

(1) fission only   (3) both fission and fusion
(2) fusion only   (4) neither fission nor fusion

_____   **a.** A very large nucleus splits into smaller pieces.

_____   **b.** The total mass before a reaction is greater than the mass after a reaction.

_____   **c.** The rate of a reaction can be safely controlled for energy generation in suitable vessels.

_____   **d.** Two small nuclei form a single larger one.

_____   **e.** Less-stable nuclei are converted to more-stable nuclei.

2. Match the reaction type on the right to the statement(s) that applies to it on the left.

_____   It requires very high temperatures.   **(a)** uncontrolled fusion

_____   It is used in nuclear reactors to make electricity.   **(b)** uncontrolled fission

_____   It occurs in the sun and other stars.   **(c)** controlled fusion

_____   It is used in atomic bombs.   **(d)** controlled fission

3. Match the component of a nuclear power plant on the right to its use on the left.

_____   limits the number of free neutrons   **(a)** moderator

_____   is used to slow down neutrons   **(b)** fuel rod

_____   drives an electric generator   **(c)** control rod

_____   provides neutrons by its fission   **(d)** shielding

_____   removes heat from the system safely   **(e)** coolant

_____   prevents escape of radiation   **(f)** turbine

4. _____ A chain reaction is any reaction in which

    (a) excess reactant is present.   **(c)** the rate is slow.

    (b) the material that starts the reaction is also a product.   **(d)** many steps are involved.

SECTION 4 continued

5. As a star ages, does the ratio of He atoms to H atoms in its composition become larger, smaller, or remain constant? Explain your answer.

_____

_____

_____

_____

6. The products of nuclear fission are variable; many possible nuclides can be created. In the feature "An Unexpected Finding," on page 702 of the text, it was noted that Meitner showed radioactive barium to be one product of fission. Following is an incomplete possible nuclear equation for the production of barium-141:

$$^{235}_{92}U + ^{1}_{0}n \rightarrow ^{141}_{56}Ba + \underline{\quad ? \quad} + 4\,^{1}_{0}n + \text{energy}$$

_____ **a.** Determine the missing fission product formed. Write the element's symbol, its atomic number, and its mass number.

_____ **b.** Is it likely that this isotope in part **a** is unstable? (Refer to **Figure 2** on page 685 of the text.)

7. Small nuclides can undergo fusion.

_____ **a.** Complete the following nuclear equation by identifying the missing term. Write the element's symbol, its atomic number, and its mass number.

$$^{3}_{1}H + ^{7}_{3}Li \rightarrow \underline{\quad ? \quad} + \text{energy}$$

b. When measured exactly, the total mass of the reactants does not add up to that of the products in the reaction represented in part **a**. Why is there a difference between the mass of the products and the mass of the reactants? Which has the greater mass, the reactants or the products?

_____

_____

_____

8. What are some current concerns regarding development of nuclear power plants?

_____

_____

_____

_____

CHAPTER 21 REVIEW
Nuclear Chemistry

MIXED REVIEW

SHORT ANSWER   Answer the following questions in the space provided.

1. The ancient alchemists dreamed of being able to turn lead into gold. By using lead-206 as the target atom of a powerful accelerator, modern chemists can attain that dream in principle. Write the nuclear equation for a one-step process that will convert $^{206}_{82}Pb$ into a nuclide of gold-79. You may use alpha particles, beta particles, positrons, or protons.

_____

2. A typical fission reaction releases 2×10^{10} kJ/mol of uranium-235, while a typical fusion reaction produces 6×10^8 kJ/mol of hydrogen-1. Which process produces more energy from 235 g of starting material? Explain your answer.

_____

_____

_____

_____

3. Write the nuclear equations for the following reactions:

  a. Carbon-12 combines with hydrogen-1 to form nitrogen-13.

_____

  b. Curium-246 combines with carbon-12 to form nobelium-254 and four neutrons.

_____

  c. Hydrogen-2 combines with hydrogen-3 to form helium-4 and a neutron.

_____

4. Write the complete nuclear equations for the following reactions:

  a. $^{91}_{42}Mo$ undergoes positron emission.

_____

  b. $^{6}_{2}He$ undergoes beta decay.

_____

  c. $^{194}_{84}Po$ undergoes alpha decay.

_____

MIXED REVIEW continued

    d. $^{129}_{55}$Cs undergoes electron capture.

_____

PROBLEMS   Write the answer on the line to the left. Show all your work in the space provided.

5. _____ It was shown in **Section 1** of the text that a mass defect of 0.030 377 amu corresponds to a binding energy of 4.54×10^{-12} J. What binding energy would a mass defect of 0.015 amu yield?

6. _____ Iodine-131 has a half-life of 8.0 days; it is used in medical treatments for thyroid conditions. Determine how many days must elapse for a 0.80 mg sample of iodine-131 in the thyroid to decay to 0.10 mg.

7. Following is an incomplete nuclear fission equation:

$$^{235}_{92}U + ^{1}_{0}n \rightarrow ^{90}_{38}Sr + ^{141}_{54}Xe + x\,^{1}_{0}n + \text{energy}$$

_____ **a.** Determine the value of x in the above equation.

_____ **b.** The strontium-90 produced in the above reaction has a half-life of 28 years. What fraction of strontium-90 still remains in the environment 84 years after it was produced in the reactor?

Name _____ Date _____ Class _____

SECTION 1

SHORT ANSWER Answer the following questions in the space provided.

1. Name two types of carbon-containing molecules that are not organic.

_____

2. _____ Carbon atoms form bonds readily with atoms of

    (a) elements other than carbon.   **(c)** both carbon and other elements.
    (b) carbon only.   **(d)** only neutral elements.

3. Explain why the following two molecules are *not* geometric isomers of one another.

```
      H   Cl                          H   H
      |   |                           |   |
  H — C — C — H                   H — C — C — H
      |   |                           |   |
      Cl  H                           Cl  Cl
```

_____

_____

_____

4. a. In the space below, draw the structural formulas for two structural isomers with the same molecular formula.

b. In the space below, draw the structural formulas for two geometric isomers with the same molecular formula.

SECTION 1 continued

5. Draw a structural formula that demonstrates the catenation of the methane molecule, CH_4.

6. Draw the structural formulas for two structural isomers of C_4H_{10}.

7. Draw the structural formula for the *cis*-isomer of $C_2H_2Cl_2$.

8. Draw the structural formula for the *trans*-isomer of $C_2H_2Cl_2$.

CHAPTER 22 REVIEW
Organic Chemistry

SECTION 2

SHORT ANSWER Answer the following questions in the space provided.

1. _____ Hydrocarbons that contain only single covalent bonds between carbon atoms are called

   (a) alkanes.   (c) alkynes.
   (b) alkenes.   (d) unsaturated.

2. _____ When the longest straight-chain in a hydrocarbon contains seven carbons, its prefix is

   (a) pent-.   (c) hept-.
   (b) hex-.   (d) oct-.

3. _____ The alkyl group with the formula —CH_2—CH_3 is called

   (a) methyl.   (c) propyl.
   (b) ethyl.   (d) butyl.

4. What is a saturated hydrocarbon?

   _____

   _____

5. Explain why the general formula for an alkane, C_nH_{2n+2}, correctly predicts hydrocarbons in a homologous series.

   _____

   _____

   _____

   _____

6. Why is the general formula for cycloalkanes, C_nH_{2n}, different from the general formula for straight-chain hydrocarbons?

   _____

   _____

SECTION 2 continued

7. Write the IUPAC name for the following structural formulas:

_____ a.

$$CH_3-CH_2 \quad CH_2-CH_3$$
$$\qquad | \qquad\qquad |$$
$$CH_3-CH_2-CH-CH-CH_2-CH_3$$

_____ b.

$$CH_3-CH_2$$
$$\qquad\qquad |$$
$$CH_3-CH_2-C-CH_2-CH_3$$
$$\qquad\qquad |$$
$$\qquad\qquad CH_3$$

_____ c.

$$CH_3-CH-CH_2-CH_2-CH-CH-CH_2-CH_3$$
$$\qquad | \qquad\qquad\qquad | \quad |$$
$$\qquad CH_3 \qquad\qquad\quad CH_3 \ CH_3$$

8. Draw the structural formula for each of the following compounds:

a. 3,4-diethyl-2-methy-1-hexene

b. 1-ethyl-2,3-dimethylbenzene

c. 5, 6-dimethyl-2-heptyne

CHAPTER 22 REVIEW
Organic Chemistry

SECTION 3

SHORT ANSWER Answer the following questions in the space provided.

1. Match the structural formulas on the right to the family name on the left.

_____ aldehyde

_____ ketone

_____ carboxylic acid

_____ amine

_____ ester

_____ alkene

(a) H—C—OH
         ‖
         O

(b)   H   H   H
         | | |
     H—C—N—C—H
         | |
         H   H

(c)   H
            |
   H—C—O—C—H
       ‖ |
       O   H

(d)   H
         |
   H—C—C=O
         |
         H   H

(e) H   H
       \   /
        C=C
       /   \
     H   H

(f)   H   O   H
      | ‖ |
  H—C—C—C—H
      | |
      H   H

2. What is the functional group in glycerol? Explain how glycerol functions in skin care products.

_____

_____

_____

3. List the halogen atoms found in alkyl halides in order of increasing atomic mass.

_____

_____

4. State the difference between aldehydes and ketones.

_____

_____

_____

SECTION 3 continued

5. _____ Which is the weaker acid, acetic acid or sulfuric acid?

6. Explain why esters are considered derivatives of carboxylic acids.

_____

_____

7. Draw structural formulas for the following compounds:

  a. 1-butanol

  b. dichlorodifluoromethane

CHAPTER 22 REVIEW
Organic Chemistry

SECTION 4

SHORT ANSWER   Answer the following questions in the space provided.

1. Match the reaction type on the left to its description on the right.

_____ substitution

(a) An atom or molecule is added to an unsaturated molecule, increasing the saturation of the molecule.

_____ addition

(b) A simple molecule is removed from adjacent atoms of a larger molecule.

_____ condensation

(c) One or more atoms replace another atom or group of atoms in a molecule.

_____ elimination

(d) Two molecules or parts of the same molecule combine.

2. Substitution reactions can require a catalyst to be feasible. The reaction represented by the following equation is heated to maximize the percent yield.

$$C_2H_6(g) + Cl_2(g) + energy \overset{\Delta}{\rightleftarrows} C_2H_5Cl(l) + HCl(g)$$

_____ **a.** Should a high or low temperature be maintained?

_____ **b.** Should a high or low pressure be used?

_____ **c.** Should the HCl gas be allowed to escape into another container?

3. Elemental bromine is a reddish-brown liquid. Hydrocarbon compounds that contain bromine are colorless. A qualitative test for carbon-carbon multiple bonds is the addition of a few drops of bromine solution to a hydrocarbon sample at room temperature and in the absence of sunlight. The bromine will either quickly lose its color or remain reddish brown.

_____ **a.** If the sample is unsaturated, what type of reaction should occur when the bromine is added under the conditions mentioned above?

_____ **b.** If the sample is saturated, what type of reaction should occur when the bromine is added under the conditions mentioned above?

_____ **c.** The reddish brown color of a bromine solution added to a hydrocarbon sample at room temperature and in the absence of sunlight quickly disappears. Is the sample a saturated or unsaturated hydrocarbon?

SECTION 4 continued

4. Two molecules of glucose, $C_6H_{12}O_6$, undergo a condensation reaction to form one molecule of sucrose, $C_{12}H_{22}O_{11}$.

   _____ **a.** How many molecules of water are formed during this condensation reaction?

   b. Write a balanced chemical equation for this condensation reaction.

   _____

5. Addition reactions with halogens tend to proceed rapidly and easily, with the two halogen atoms bonding to the carbon atoms connected by the multiple bond. Thus, only one isomeric product forms.

   a. Write an equation showing the structural formulas for the reaction of Br_2 with 1-butene.

   b. Name the product.

   _____

6. Identify each of the following substances as either a natural or a synthetic polymer.

   _____ **a.** cellulose

   _____ **b.** nylon

   _____ **c.** proteins

7. The text gives several abbreviations commonly used in describing plastics or polymers. For each of the following abbreviations, give the full term and one common household usage.

   a. HDPE

   _____

   b. LDPE

   _____

   c. cPE

   _____

8. Explain why an alkane cannot be used as the monomer of an addition polymer.

   _____

   _____

   _____

CHAPTER **22** REVIEW
Organic Chemistry

MIXED REVIEW

SHORT ANSWER   Answer the following questions in the space provided.

1. _____ A saturated organic compound

      (a) contains all single bonds.
      (b) contains at least one double or triple bond.
      (c) contains only carbon and hydrogen atoms.
      (d) is quite soluble in water.

2. Arrange the following in order of increasing boiling point:

_____ **a.** ethane

_____ **b.** pentane

_____ **c.** heptadecane

3. Recall that isomers in organic chemistry have identical molecular formulas but different structures and IUPAC names.

_____ **a.** Two isomers must have the same molar mass. True or False?

_____ **b.** Two isomers must have the same boiling point. True or False?

4. Explain why hydrocarbons with only single bonds cannot form geometric isomers.

_____

_____

_____

5. Write the IUPAC name for the following structural formulas:

a. _____

$$CH_3-CH_2 \quad CH_2-CH_3$$
$$CH_3-CH-CH-CH-CH_2-CH_3$$
$$\qquad\qquad\qquad CH_3$$

b. _____

$$CH_3$$
$$CH_3-CH-CH=CH_2$$

_____ **c.**

$$
\begin{array}{cccc}
Cl & Cl & H & Cl \\
| & | & | & | \\
H-C-C-C-C-H \\
| & | & | & | \\
H & H & H & H
\end{array}
$$

6. Draw the structural formula for each of the following compounds:

   a. 1,2,4-trimethylcyclohexane

   b. 3-methyl-1-pentyne

7. Each of the following names implies a structure but is not a correct IUPAC name. For each example, draw the implied structural formula and write the correct IUPAC name.

   a. 3-bromopropane

   b. 3, 4-dichloro-4-pentene

8. Match the general formula on the right to the corresponding family name on the left.

_____ carboxylic acid   **(a)** $R-X$

_____ ester

_____ alcohol

(b) $R-\overset{\overset{\displaystyle O}{\|}}{C}-H$

_____ ether

_____ alkyl halide   **(c)** $R-O-R'$

_____ amine   **(d)** $R-OH$

_____ aldehyde

_____ ketone   **(e)** $R-\overset{\overset{\displaystyle O}{\|}}{C}-OH$

(f) $R-\overset{\overset{\displaystyle }{\underset{\underset{\displaystyle R'}{|}}{N}}}-R''$

(g) $R-\overset{\overset{\displaystyle O}{\|}}{C}-R'$

(h) $R-\overset{\overset{\displaystyle O}{\|}}{C}-O-R'$

CHAPTER 23 REVIEW
Biological Chemistry

SECTION 1

SHORT ANSWER   Answer the following questions in the space provided.

1. _____ Lactose and sucrose are both examples of

  (a) lipids.
  (b) monosaccharides.
  (c) disaccharides.
  (d) proteins.

2. _____ Carbohydrates made up of long chains of glucose units are called

  (a) monosaccharides.
  (b) disaccharides.
  (c) polysaccharides.
  (d) simple sugars.

3. _____ The disaccharide that is commonly known as table sugar is

  (a) lactose.
  (b) fructose.
  (c) sucrose.
  (d) maltose.

4. _____ The polysaccharide that plants use for storing energy is

  (a) starch.
  (b) glycerol.
  (c) cellulose.
  (d) glycogen.

5. _____ Many animals store carbohydrates in the form of

  (a) starch.
  (b) glycogen.
  (c) cellulose.
  (d) glycerol.

6. _____ Which class of biomolecules includes fats, oils, waxes, steroids, and cholesterol?

  (a) starches
  (b) monosaccharides
  (c) disaccharides
  (d) lipids

SECTION 1 continued

7. Relate the structure of carbohydrates to their role in biological systems.

_____

_____

_____

_____

_____

_____

_____

8. What is a condensation reaction, what is a hydrolysis reaction, and how do they differ?

_____

_____

_____

_____

_____

_____

9. Why can cows digest cellulose, while humans cannot?

_____

_____

_____

_____

_____

10. Describe how phospholipids are arranged in the cell membrane.

_____

_____

_____

_____

_____

_____

_____

CHAPTER 23 REVIEW
Biological Chemistry

SECTION 2

SHORT ANSWER Answer the following questions in the space provided.

1. _____ Proteins are polypeptides made of many

  (a) lipids.
  (b) carbohydrates.
  (c) starches.
  (d) amino acids.

2. _____ The side chains of amino acids may contain

  (a) acidic and basic groups.
  (b) polar groups.
  (c) nonpolar groups.
  (d) All of the above

3. _____ The amino acid sequence of a polypeptide chain is its

  (a) primary structure.
  (b) secondary structure.
  (c) tertiary structure.
  (d) quaternary structure.

4. _____ The secondary structure of a protein that is shaped like a coil, with hydrogen bonds that form along a single segment of peptide, is

  (a) a looped structure.
  (b) the active site.
  (c) an alpha helix.
  (d) a beta pleated sheet.

5. According to the text, which amino acid(s) contains a side chain

  _____ **a.** in which molecules form covalent disulfide bridges with each other?

  _____ **b.** that is hydrophobic?

  _____ **c.** that forms hydrogen bonds?

  _____ **d.** that is basic?

6. Identify each protein structure level described below.

_____ **a.** may involve hydrogen bonds, salt bridges, and disulfide bonds that determine a protein's three-dimensional structure

_____ **b.** is determined by the interaction of several polypeptides coming together

_____ **c.** is the amino acid sequence of a protein

7. How many different tripeptides can be formed from one molecule of glycine and two molecules of valine? Draw the isomers using their three-letter codes.

_____

_____

8. What are the functions of fibrous proteins?

_____

_____

_____

_____

9. How do enzymes work?

_____

_____

_____

_____

_____

CHAPTER 23 REVIEW
Biological Chemistry

SECTION 3

SHORT ANSWER   Answer the following questions in the space provided.

1. _____ The primary energy exchange in the body is the cycle between

    (a) amino acids and proteins.
    (b) ATP and ADP.
    (c) lipids and carbohydrates.
    (d) DNA and RNA.

2. _____ In which equation is the hydrolysis reaction of ATP represented?

    (a) $ADP^{3-}(aq) + H_2O(l) \rightarrow ATP^{4-}(aq) + H_2PO^{4-}(aq)$
    (b) $ATP^{4-}(aq) + H_2PO^{4-}(aq) \rightarrow ADP^{3-}(aq) + H_2O(l)$
    (c) $ATP^{4-}(aq) + H_2O(l) \rightarrow ADP^{3-}(aq) + H_2PO^{4-}(aq)$
    (d) None of the above

3. _____ In the citric acid cycle,

    (a) CO_2 and ATP are formed.
    (b) food is digested.
    (c) glucose is formed.
    (d) DNA is replicated.

4. _____ Animals can produce ATP molecules in

    (a) photosynthesis.
    (b) the Krebs cycle.
    (c) peptide synthesis.
    (d) DNA replication.

5. _____ In glucogenesis, glucose is synthesized from

    (a) sucrose and fructose.
    (b) water and amino acids.
    (c) lactate, pyruvate, glycerol, and amino acids.
    (d) DNA and RNA.

Name _____ Date _____ Class _____

6. Identify each function as that of *autotrophs* or of *heterotrophs*.

a. _____ synthesize carbon-containing biomolecules from H_2O and CO_2

b. _____ absorb solar energy, which is converted into ATP

c. _____ obtain energy by consuming plants or animals

7. Identify each as either a *catabolic process* or an *anabolic process*.

a. Synthesis of protein molecules is a(n) _____.

b. A(n) _____ releases energy.

c. Digestion is a(n) _____.

d. A(n) _____ requires energy.

8. How do plants use photosynthesis to gather energy?

_____
_____
_____
_____
_____
_____

9. Explain how animals indirectly gather energy from the sun.

_____
_____
_____
_____
_____
_____

MODERN CHEMISTRY

CHAPTER 23 REVIEW
Biological Chemistry

SECTION 4

SHORT ANSWER Answer the following questions in the space provided.

1. Complete the following statements with *DNA* or *RNA*.

   a. _____ is most often found in the form of a double helix.

   b. _____ contains ribose as its sugar unit.

   c. _____ is most often single stranded.

   d. _____ is directly responsible for the synthesis of proteins.

2. Complete the following statements with the name of the correct base. More than one answer may be used.

   a. _____ contains a six-membered ring called a pyrimidine.

   b. _____ is the complementary base of A in RNA.

   c. _____ contains a five-membered ring called a purine.

   d. _____ is the complementary base of G in DNA.

3. A segment of DNA has the base sequence TAC TTT TCG AAG AGT ATT.

   a. What is the base sequence in a complementary strand of RNA?

   _____

   b. What is the base sequence in a complementary strand of DNA?

   _____

4. A segment of DNA has the base sequence TAC CTT ACA GAT TGT ACT.

   a. What is the base sequence in a complementary strand of RNA?

   _____

   b. What is the base sequence in a complementary strand of DNA?

   _____

SECTION 4 continued

5. Explain why, in DNA, pairing exists only between A and T and between C and G?

_____

_____

_____

_____

_____

_____

6. What is cloning, and how has it been accomplished in mammals?

_____

_____

_____

_____

_____

_____

CHAPTER 23 REVIEW
Biological Chemistry

MIXED REVIEW

SHORT ANSWER   Answer the following questions in the space provided.

1. Use *carbohydrate(s), lipid(s), protein(s),* or *DNA* to complete the following statements.

    a. _____ is the largest molecule found in cells.

    b. _____ are the major component in a cell membrane.

    c. _____ provides most of the energy that is available in plant-derived food.

    d. _____ gets its name from the Greek word meaning "of first importance."

2. Describe four different kinds of interactions between side chains on a polypeptide molecule that help to make the shape that a protein takes.

_____

_____

_____

_____

_____

_____

3. How does DNA replicate itself?

_____

_____

_____

_____

_____

_____

4. Draw the reaction of ATP hydrolysis to ADP, indicating the free energy.

5. For the following peptide molecule, identify the peptide linkages, the amino groups, and the carboxyl groups.

6. A segment of RNA has the base sequence UAG CCU AAG CGA UAC GGC ACG.

a. What is the base sequence in a complementary strand of RNA?

_____

b. What is the base sequence in the complementary strand of DNA?

_____

7. Draw the condensation reaction of two molecules of glucose.glucose.

glucose   glucose